ENCOUNTERING AF

To Edith, Stanley and Rach

Encountering Affect

Capacities, Apparatuses, Conditions

BEN ANDERSON
Durham University, UK

LONDON AND NEW YORK

First published 2014 by Ashgate Publishing

Published 2016 by Routledge
2 Park Square, Milton Park, Abingdon, Oxfordshire OX14 4RN
711 Third Avenue, New York, NY 10017, USA

First issued in paperback 2016

Routledge is an imprint of the Taylor & Francis Group, an informa business

British Library Cataloguing in Publication Data
A catalogue record for this book is available from the British Library

The Library of Congress has cataloged the printed edition as follows:
Anderson, Ben, 1972-
Encountering affect : capacities, apparatuses, conditions / by Ben Anderson.
pages cm
Includes bibliographical references and index.
ISBN 978-0-7546-7024-7 (hardback : alk. paper) -- ISBN 978-1-4724-3777-8 (ebook) --
ISBN 978-1-4724-3778-5 (epub) 1. Affect (Psychology) 2. Emotions. I. Title.
BF175.5.A35A53 2014
152.4--dc23

2013049438

ISBN 13: 978-1-138-24848-9 (pbk)
ISBN 13: 978-0-7546-7024-7 (hbk)

Contents

Acknowledgements

My first thanks must go to friends who, as well as a lot more, kept reminding me that I should be writing this book rather than doing something else, especially Peter Adey, Stuart Elden and Colin McFarlane. I was usually doing something else, but I appreciated that they asked where the book was. The book owes much to conversations over the years with Paul Harrison, Angharad Closs-Stephens, Adam Holden and other friends at Durham who have helped make the geography department such an inspiring and supportive place to work over the last 10 years, in particular Ash Amin, Louise Amoore, Andrew Baldwin, David Bissell, Mike Crang, Jonny Darling, Bethan Evans, Nicky Gregson, Jenny Laws, Patrick Murphy, Joe Painter, Gordon Macleod, Ruth Raynor, Rob Shaw, Dan Swanton and Helen Wilson. The interest in hope, boredom and affect started life in doctoral work over 10 years ago. I still owe a debt to Gill Valentine and Peter Jackson, who supervised my PhD in a way that allowed me to follow my interests, even if they weren't ones that they necessarily shared. I was also lucky to be doing a PhD around the same time as Louisa Cadman, who was one of the people responsible for introducing me to a different Foucault. He probably has no idea about this, but many of my interests here were first sparked by seminars at Sheffield by Nick Bingham on actor-network theory. I have been lucky enough to supervise a great bunch of PhD students, some of whom share an interest in affect, and are developing work on affect and emotion far beyond this book; Simon Beer, Ladan Cockshut, Matt Finn, Peter Forman, Rachel Gordon, Emily Jackson, Charlotte Lee, Patrick Murphy, Eduardo Neve and Nathaniel O'Grady. The book has also benefited hugely from conversations with a large number of people who have passed through Durham over the years, have commented on one or more of its chapters or the manuscript/original proposal, or have just been part of a supportive milieu, including Jane Bennett, William E. Connolly, Joyce Davidson, J-D Dewsbury, Mick Dillon, Jason Lim, Derek McCormack, Greg Seigworth, Nigel Thrift, Keith Woodward and John Wylie. Thanks should also go to Valerie Rose for her support for the initial idea at a time when affect was very much a marginal interest and, most of all, for the gracious way she accepted my frequent apologies at various conferences for the book's lateness. Finally, thanks to Rach, Edith and now Stanley for the life we live together. Part of the reason why writing this book has taken a while is that being with them has been so much more enjoyable, important and fun.

Section 1.1 includes edited excerpts from Anderson, B. (2013) Affect and Emotion. In: Johnson, N., Schein, R. and Winders, J. (eds) *The Wiley-Blackwell*

Companion to Cultural Geography. London: Wiley-Blackwell, 452–64, reproduced with permission of Wiley-Blackwell.

Section 2.4 includes an edited version of part of Anderson, B. (2010) Modulating the Excess of Affect: Morale in a State of Total War. In: Gregg, M. and Seigworth, G. *The Affect and Cultural Theory Reader*. Durham, NC and London: Duke University Press. ©2010, Duke University Press. All rights reserved. Reprinted by permission of the publisher. www.dukeupress.edu. The first three pages of Section 2.3 include edited excerpts from Anderson, B. and Adey, P. (2011) Affect and security: Exercising emergency in 'UK civil contingencies'. *Environment and Planning D: Society and Space* 29, 1092–1109, reproduced with permission of Pion Ltd, London (www.pion.co.uk and www.envplan.com)

Section 4.4 includes an edited version of part of Anderson, B. (2006) Becoming and being hopeful: towards a theory of affect. *Environment and Planning D: Society and Space* 24: 733–52, reproduced with permission of Pion Ltd, London (www.pion.co.uk and www.envplan.com)

Section 6.1, 6.2 and 6.4 include brief edited excerpts from Anderson, B. (2009) Affective atmospheres. *Emotion, Space and Society* 2: 77–81, reproduced with permission of Elsevier.

Chapter 1
Affective Life

1.1 Hope and Other Affects

This book is about precarity, optimism, emergency, pressure, debility-dependency-dread, morale, boredom, urgency and greed. It is about the sensibilities, concepts and theories needed to understand how these and other affects relate to and become part of social-spatial relations. And it is about the connections between affective life and processes of mediation. It is also about hope.

1.1.1 Events of Hope

For 17 days families, friends and then a global media audience waited. Trapped in the emergency shelter of a collapsed mine 2,300 ft into the earth, 33 Chilean miners waited to be rescued. During this period, hope was kept alive and lost, given and received. The first images of the men gave hope to the families who waited in the self-titled Camp Hope as the ordeal went on. Hope was described by NASA experts in human confinement as a resource that would enable the miners to cope deep in the earth. Here is how Carola Narvaez, the wife of Raul Bustos one of the trapped miners, expresses her hope in the context of a previous disaster, an earthquake, they had survived together:

> In the earthquake we just had to keep on living, we had our lives … this is the same. It is producing much anguish, isolation, fear. But we're alive. My husband is alive down in that mine and we will have another happy ending.[1]

In his inaugural address to the University of Tübingen in 1961 the Marxist process philosopher Ernst Bloch, speaking in the shadow of Nazi Germany, asked a simple question about the event of hope: can hope be disappointed? His answer was yes, to be hope it must be disappointable. Hopes and hoping open up a point of contingency in the here and now. Indeed:

> [h]ope must be unconditionally disappointable … because it is open in a forward direction, in a future-orientated direction; it does not address itself to that which already exists. For this reason, hope – while actually in a state of suspension – is

1 Available at: http://www.telegraph.co.uk/news/worldnews/southamerica/chile/7969012/Wife-of-Chilean-miner-tells-how-she-survive-earthquake-six-months-ago.html (last accessed 10 July 2012).

> committed to change rather than repetition, and what is more, incorporates the element of chance, without which there can be nothing new (Bloch 1998: 341).

1.1.2 Atmospheres of Hope

A new American president speaks at his inauguration. Employing the prophetic voice of the Black American church, and revitalising the future-orientated promise of the American dream, the audience regularly interrupt President Barack Obama's speech with a refrain: 'yes we can, yes we can'. Originally a slogan of Latin American and US–Mexican borderland unionism (Saldanha 2010), the refrain energises the crowd who call it back to Obama. Near the start of the speech, amid the cheers, cries of agreement and tears of the crowd, Obama evokes a moment of hope amid danger, a moment of promise amid the tangible and less tangible signs of crisis and defeat:

> Homes have been lost, jobs shed, businesses shuttered. Our health care is too costly, our schools fail too many, and each day brings further evidence that the ways we use energy strengthen our adversaries and threaten our planet.
>
> These are the indicators of crisis, subject to data and statistics. Less measurable, but no less profound, is a sapping of confidence across our land; a nagging fear that America's decline is inevitable, that the next generation must lower its sights.
>
> Today I say to you that the challenges we face are real, they are serious and they are many. They will not be met easily or in a short span of time. But know this America: They will be met.

As Obama delivers the defiant line 'They will be met' the crowd cheers. Perhaps relieved that President George W. Bush is gone, perhaps overjoyed by the occasion, perhaps delighted to see a black American finally become president, the crowd will be characterised by participants and commentators above all else as hopeful. Perhaps this atmosphere is replicated in the many sites that host inauguration parties and events. Churches, theatres, cinemas, restaurants and coffee shops all show the inauguration live and are animated by cheers and tears as atmospheres are formed and deform, catalysed by Obama's reassertion of the American dream and the democratic promise.

1.1.3 Knowing Hope

Every month since 1967 The Conference Board,[2] a US business membership and research association, has released a monthly 'consumer confidence index'.

---

2 The Conference Board is a business membership and research association founded in 1916. Its website describes its 'unique mission' in the following terms: 'To provide the world's leading organizations with the practical knowledge they need to improve their

Designed to measure the degree of optimism about the economy as expressed in aggregate patterns of spending and saving, the index establishes trends in consumers' affectively imbued relation to the future. Writing in November 2010, a research advisor at the Federal Reserve Bank of San Francisco – Sylvain Leduc – notes that economists are worried about the US economy:

> Indicators of consumer confidence have been at depressed levels in recent months. Business sentiment is also low, reflecting uncertainty about US fiscal policy and the perception that economic weakness may be prolonged. This lack of confidence raises the risk that pessimism can become entrenched and self-reinforcing, further dampening the nascent recovery (Leduc 2010: 22 November).

Influenced by Keynes's (1936) comments about the role of 'animal spirits' in driving economic activity, economists have long debated whether confidence is an independent economic variable and thus a 'business cycle driver'. Although there is no consensus as to whether confidence is cause or effect, optimism regarding the economy is nevertheless measured, presented graphically, and trends identified and discussed.[3] Through household surveys focused on present and future spending and saving, changes at the aggregate level of the 'degree of optimism' of a population come to be known and tracked. For example, The Conference Board's consumer confidence index is based on a survey of 5,000 US households. Questions asked include: What are your expectations for the general employment situation in six months (better: the same: worse)? What are your expectations for your personal income in six months (higher: the same: lower)? On the basis of the consumer confidence index and other surveys, action may be taken to respond to or attempt to change consumer optimism or pessimism. Markets may react, monetary policy may be changed or businesses may boost investment.

The three examples are on first impression about a similar relation to the future – hope or optimism – albeit in diverse contexts: disaster relief, political speech and macroeconomic policy. I begin this book with the three scenes because they each exemplify a specific way in which affective life takes place and is organised. In

---

performance and better serve society' (see www.conference-board.org/about/, accessed 10 February 2012).

3 To quote Keynes (1936: 161–2) on the relation between economic fluctuations, uncertainty and shifts in mood (including consumer confidence): 'Even apart from the instability due to speculation, there is the instability due to the characteristic of human nature that a large proportion of our positive activities depend on spontaneous optimism rather than mathematical expectations, whether moral or hedonistic or economic. Most, probably, of our decisions to do something positive, the full consequences of which will be drawn out over many days to come, can only be taken as the result of animal spirits – a spontaneous urge to action rather than inaction, and not as the outcome of a weighted average of quantitative benefits multiplied by quantitative probabilities'.

the example of the trapped miners, acts of hope open up a moment of difference in the context of a shared situation of misery and suffering. Hope is kept alive, if only just, and the suffering that marks the present is disrupted, if only momentarily. In the second example, hope is akin to an affective atmosphere, simultaneously absent and present, material and immaterial. Passing between bodies, a sense of possibility appears to infuse President Obama's inauguration as crowds cry and cheer. The sense of possibility became a structure of feeling: pressuring and limiting how Obama's subsequent actions were related to, later folding into disappointment. In the final example, hope is named, known and rendered actionable. Relations to the future are translated into an index. Techniques are deployed by economists to know optimism and macroeconomic policy may change in response to aggregate fluctuations in collective mood. In this case, collective optimism is the target and focus for an intervention in the hope of increasing aggregate demand in an economy teetering on the brink of catastrophe and faltering in the midst of a crisis.

How might geographers and other social scientists encounter these and other ways in which affective life happens? How to understand the ways in which affective life comes to be temporarily organised in relation to social, political and other processes, forms and forces? In addressing these questions the book outlines a distinctive analytics of affect orientated to three ways of encountering and understanding affect. The three ways are exemplified by the scenes of hope with which I began the book. Affect is: an *object-target* in the example of consumer confidence; a *bodily capacity* emergent from encounters in the rescue of the miners; and a *collective condition* in Obama's inauguration. Understanding the ongoing mediation, organisation and surprise of affective life involves addressing questions specific to each understanding of affect: how is affective life an object-target for specific and multiple forms of power?; how do bodily capacities form in the midst of the encounters that make up living?; and how do collective affects take place so that they become part of the conditions for life? The book explores some of the concepts, sensibilities and techniques through which we might address these questions. To this end, the book gathers together insights from a range of theories of affect. It does not comprehensively review every theory of affect and emotion in human geography, let alone in the social sciences and humanities.[4] Such a task is probably now impossible, given the sheer range and variety of affect theories. Instead, I explore a number of concepts that have struck me and stayed with me. Concepts that have offered me particular ways of understanding the imbrication of affect with everything from the War on Terror through to listening

4 For reviews of work on emotions in the social science see Lupton (1998) and Williams (2001). Generally, reviews parcel work on emotions into a number of recognised approaches, principally some variant of Lupton's (1998) distinction between: emotions as inherent (with appraisal theories being a subcategory) and emotions as sociocultural constructions (with phenomenological, poststructuralist and psychoanalytical approaches being distinguished from one another). In geography, Bondi et al. (2005) provide a careful overview of work on emotion/affect.

to music. In doing so I outline one specific way of relating to and understanding affective life that coexists alongside other theories of affect and emotion. But first let us step back and examine why considering affect matters…

1.2 Promises, Imperatives

In an essay first published in 1995, Brian Massumi (2002a: 27) articulated a 'growing feeling' that affect was central to understanding what he termed, after Ernest Mandel (1978), 'our information- and image-based late capitalist culture'. Writing 17 or so years later, it seems as though this 'growing feeling' has crossed a threshold to become something close to a starting point for recent theory. Patricia Clough (with Jean Halley) (2007) has identified, rightly I think, that an 'affective turn' has occurred across a range of social sciences and humanities (see also Gregg and Seigworth 2010; Blackman and Venn 2010; Greco and Stenner 2008). Naming a real world referent, a concept and something close to an ethos, the term 'affect' has been used to describe a heterogeneous range of phenomena that are taken to be part of life: background moods such as depression, moments of intense and focused involvement such as euphoria, immediate visceral responses of shame or hate, shared atmospheres of hope or panic, eruptions of passion, lifelong dedications of love, fleeting feelings of boredom, societal moods such as anxiety or fear, neurological bodily transitions such as a feeling of aliveness, waves of feeling … amongst much else. In human geography alone the term has been used to understand a wonderfully diverse range of geographies; fathering (Aitken 2009), popular geopolitics (Carter and McCormack 2006), landscape relations (Wylie 2009), new forms of work (Woodward and Lea 2010), race and racism (Lim 2010; Swanton 2010), alcohol (Jayne, Valentine and Holloway 2010), obesity (Evans 2010), dance (McCormack 2003), war and violences (Ó Tuathail 2003), therapeutic landscapes (Conradson 2010; Lea 2008), animals and other non-humans (Greenhough and Roe 2010; Roe 2006) and technological life (Ash 2012; Kinsley 2010), to name but some.

Given this diversity, I begin with a simple affirmation, one that is at the heart of Eve Kosofsky Sedgwick's (2003) influential discussion of affect. In the conclusion to her essay on shame and the novelist Henry James, Sedgwick (62) describes one affect – shame – as a 'kind of free radical'. By which she means that shame '[a]ttaches to and permanently intensifies or alters the meaning of – of almost anything' (62). Always insisting on the plural – affects rather than the singular affect – Sedgwick's writings bear witness to and express the combinatorial complexity of shame, love, paranoia and other affects.[5] Her deceptively simple

---

5 Probyn (2005) makes the same call in her discussion of shame: we should develop concepts, sensibilities and methods that enable us to attend to the specificities of affects rather than continue to invoke the properties or capacities of a mysterious substance called affect in general.

starting point is one shared with the recent work cited above on affect in geography; the freedom of the affects to combine with more or less any aspect of life. This means that there can never be a carefully bounded affectual or emotional geography separate from other geographies. Summarising Silvan Tomkins, Sedgwick opens up work on affect since:

> Affects can be, and are, attached to things, people, ideas, sensations, relations, activities, ambitions, institutions, and any number of other things, including other affects (Sedgwick 2003: 19).

In thinking of affects as 'free radicals' coursing through life, Sedgwick's work invokes the dynamism of affective life, whilst never forgetting that affects do become attached to ... almost anything. Affects are constantly infusing embodied practices, resonating with discourses, coalescing around images, becoming part of institutions, animating political violences, catalysing political communities, and being known and intervened in, amongst much else. Or to return to the three examples I started with, affects such as hope may become part of global media events, forms of political speech or macroeconomic policy. Cutting across the separate domains we habitually organise the world into, affects are not the special property of any one domain of life (economic, cultural, and so on) or functionally distinct sector (law, medicine, art, etc.).

However, the term 'affect' is curious. Unlike emotion, mood, feeling and passion, affect is not part of the standard Euro-American lexicon. Its origins are in the third person description of affective states by psychologists, although its genealogy or history remains to be written (Ngai 2005; Dixon 2003). Its current popularisation cuts across a range of theories, each of which varies in how they conceptualise what affect is and does, how the relations between affective life, space and mediation should be conceptualised and what form an affective politics can and should take. There is also a range of notable antecedents that any work on affect is indebted to; principally feminist work that troubled the distinction between reason and its opposites, cut the naturalised link between women and emotions, and showed how and why emotions matter (Lloyd 1984; Rose 1993; Anderson and Smith 2001). The affective turn is not new. Its condition is the dictum that the 'personal is political', and it is enabled by a long tradition of feminist scholarship on emotional life. This means that any book entitled *Encountering Affect* has to deal with what might seem to be a paradox. On the one hand, a deeply ingrained Euro-American version of emotion assumes that emotions, affect, feeling and other modalities cannot be directly known. Something about the class of phenomena is assumed to exceed deliberative thought.[6] On the other

---

6 See Griffiths (1997) for an in-depth discussion of how the assumption that emotion exceeds deliberative thought is bound up with the invention of the category of emotion. In this way, recent 'non-representational' work has a series of partial connections with a long history of wondering about whether emotional life can be represented, a problem that

hand, numerous theories of affect can now be found throughout the humanities and social sciences, at the same time as diagnoses of what Greco and Stenner (2008: 2–5) call 'The Affective Society' multiply. Invocations of unthinkability have been accompanied by a proliferation of attempts to name the unnameable, to think the unthinkable, to represent what is supposedly, from some perspectives but by no means all, non-representational.[7] So the problem is no longer that emotions, affects and feelings have been downplayed, silenced or marginalised, as feminist work on emotion first rightly identified and insisted (Lloyd 1984). Rather, there is now an extraordinary proliferation of versions of what affect is and does (see Thrift 2004a; Seigworth and Gregg 2010 for summaries), a return to marginalised or forgotten approaches to affective life (Blackman 2012) and intense differences around the question of what affect is and is not (see Hemmings 2005).

Some of the differences between affect theories will be discussed in the book. For now, though, it is important to note that it is not enough simply to invoke the term 'affect' or 'emotion' for any 'non-rational' phenomenon. Different uses of the terms come freighted with more or less worked through assumptions about life, processes of mediation and how they interrelate. Nevertheless, what is shared across diverse affect theories is a sense of urgency, the sense that understanding the dynamics of affective life matters for how geography relates to life and living. The reasons given are various and contradictory: spaces and places are made through affect (Bondi, Davidson and Smith 2005); affect and thinking are always-already imbricated with one another (Connolly 2002); affects 'stick' to bodies and as such attach people to inequalities (Ahmed 2004; Boler 1999); it is through affects that subjects are constituted by and constitute worlds (Davidson 2003; Wylie 2006); representations function affectively (Latham and McCormack 2009); it is at the level of affect that the real effects of forms of power are felt and lived (Thrift 2004b); and affects open up thinking to the dynamics of non-organic life (Clough 2008), to name but some of the reasons why the 'affective turn' has commanded such attention. The promise is of a worldly geography engaged with life, one that pays close attention to the subtle, elusive, dynamics of everyday living and touches the textures of social life. When gathered together, what these promises suggest is that the turn to affect is an injunction to orientate inquiry to life and living in all their richness. In learning to attend to the vagaries of affective life, the techniques and sensibilities that compose human geography and the types of politics that animate the discipline might change.

But if the 'affective turn' seems to promise much, it has also been framed as an imperative if geography is to learn to respond to how contemporary forms of power, and their specific violences, work on and through affect. Understanding

---

connected nineteenth-century romantics and early political economy, for example (see also Gallagher 2006).

7 Here I am referencing Foucault's (1978) hypothesis on the functional effects of the 'repressive hypothesis' to highlight the relation between invocations of unknowability or unspeakability and attempts to know and speak about a phenomenon.

affect might promise a more worldly geography, but it is simultaneously an imperative if geography is to remain relevant and engaged with this world and its threats and promises. The imperative to understand how forms of power function affectively also emerges from diverse sources, some of which involve diagnosing the contemporary condition, others of which involve broader claims about the relation between affect, the political and life. The affective turn emerges from: a concern with the intimate textures of everyday life and the marginalising or silencing of specific experiences (often gendered or raced) (Probyn 2005; Ahmed 2004); from a nascent recognition that affect is formulated and transmitted in biopolitical formations addressed to 'affect itself' (Hardt and Negri 2004; Adey 2010; Clough 2004); from an acknowledgement of the emergence of new ways of tracking and knowing affectivity, including neuroscience and behaviourism, that stress the interimplication of affect and cognition (Jones, Pykett and Whitehead 2011); from changing ways of governing that address persons and collectives as affective beings and affect structures (Isin 2004; Anderson 2007); and in relation to claims about the centrality of affective or emotional labour (as one subset of immaterial labour) to post-Fordist modes of generating value (Lazzarato 1996; Hochschild 1983; 2012). The urgency of the 'affective turn' follows from a simple claim: forms of power work through affective life. Whether or not this is a transformation unique to 'late capitalist culture' is beside the point. Understanding how power functions in the early twenty-first century requires that we trace how power operates through affect and how affective life is imbued with relations of power, without reducing affective life to power's effect.

Gathered together these promises and imperatives suggest that developing a vocabulary specific to affect is a pressing challenge for contemporary geography. It is at this point, then, that differences between affect theories begin to matter. Because whilst not slavishly determining a politics, different conceptualisations of what affect is and does have consequences for understanding how life is ordered and patterned, forms of mediation work and change may happen. For this reason, I want to turn to offer an initial answer to a simple question – what is affect? – by way of an example of greed.

1.3 What is Affect?

> But in its blind unrestrainable passion, its were-wolf hunger for surplus-labour, capital oversteps not only the moral, but even the merely physical maximum bounds of the working day. It usurps the time for growth, development and healthy maintenance of the body. It steals the time required for the consumption of fresh air and sunlight. It higgles over a meal-time, incorporating it where possible with the process of production itself, so that food is given to the labourer as to a mere means of production, as coal is supplied to the boiler, grease and oil to the machinery. It reduces the sound sleep needed for restoration, reparation, refreshment of the bodily powers to just so many hours of torpor as

> the revival of an organism, absolutely exhausted, renders essential. It is not the normal maintenance of the labour-power which is to determine the limits of the working-day; it is the greatest possible daily expenditure of labour-power, no matter how diseased, compulsory, and painful it may be which is to determine the limits of the labourer's period of repose. Capital cares nothing for the length of life of labour-power. All that concerns it is simply and solely the maximum of labour-power, that can be rendered fluent in a working-day (Marx 1995: 163).

In the chapter on 'The Working Day' in Volume 1 of *Capital*, Marx describes the affects of capital and the affects of labouring bodies. The characteristic affects of capital are greed and a lack of care. Faced with capital's 'blind unrestrainable passion', workers are characterised in terms of exhaustion and fatigue. Let us pause and draw out an initial definition of affect from Marx's description of the affects of capital, the affects of labour-power, and the affects of a relation of exploitation.[8] It is also the vaguest and most general definition. Affect is a body's 'capacity to affect and be affected', where a body can in principle be anything.[9] So in the above passage affects include the 'were-wolf like' passion of capital and its lack of care, the greed of the capitalist, and the pain and tiredness of the worker and their exhausted sleep. The ordering of value-creating activities, crisis tendencies and exploitative relations given the name capitalism is made through these and other affects. Greed coexists alongside the roiling waves of panic and confidence that appear to be so integral to the creation and destruction of value in financial crises, for example.

There are two important features of this general definition, both of which lead to initial orientations for work on affective life. First, affect is two-sided. It consists of bodily capacities to affect *and* to be affected that emerge and develop in concert. For example, exhaustion both follows from a worker's position in a process of production and limits what a body can do. This initial definition has one important consequence. Straight away a body is always imbricated in a set of relations that extend beyond it and constitute it. Capacities are always collectively formed. This leads to the first question for work on affect. How are bodies formed through relations that extend beyond them and how do bodily capacities express and become part of those relations?[10] To go back to the example we could ask; how does the exhaustion of workers express their participation in what Marx later calls

8 For a compelling analysis of the affect and work that finds inspiration in Marx see Woodward and Lea (2010).

9 The point is from Deleuze's (1988a) reading of Spinoza and will be elaborated on in Chapter 4. A body can be anything; landscapes that emanate a specific atmosphere, rooms that seem imbued with a hazy feeling, enraged crowds, interesting buildings, loving couples.

10 One critique of work on affect in geography has been that it privileges the individual body, thus somehow resonating with or reproducing a link between emotion and individualism (see Saldanha 2005; Thien 2005). This critique downplays the foundational point that affects are always collective because they are constituted in and through relations.

the storms and stresses of production? Second, affect pertains to capacities rather than existing properties of the body. Affects are about what a body may be able to do in any given situation, in addition to what it currently is doing and has done. Because capacities are dependent on other bodies, they can never be exhaustively specified in advance. Here we reach Gilles Deleuze's much-quoted speculative affirmation of openness, an affirmation that follows his encounter with the joyful writings of Baruch Spinoza and has been the rallying cry of much work on affect:

> you do not know beforehand what good or bad you are capable of; you do not know beforehand what a body or mind can do, in a given encounter, a given arrangement, a given combination (Deleuze 1988a: 125).

The worker's body is not exhausted by its participation in a process of production, to go back to Marx's analysis. If it was it would have no potential to be otherwise. Indeed it is precisely an affirmation of the capacities of workers to exceed their subordination to capital that animates Marxist analysis. This means that although mediated by their relations a body's capacities are never determined by them.

Even though this first definition of affect might appear to be vague, it provides an important starting point that orientates inquiry to affect as an expression, reflection and enactment of specific relations within some form of relational configuration.[11] However, it is not enough simply to claim that affects *are* relational, and/or are emergent *from* relations and/or take place *in* relations.[12] Understood as an analytic position, relationality is a claim that can be and has been made about any and all the phenomena human geography deals with (Anderson and Harrison 2010). Whilst I do not disagree with the basic proposition, arguing that entities are 'relationally constituted' has become automatic, a habit to be mastered and repeated. Little more than the most basic starting point, it tells us nothing specific about different affects

---

Rather than an opposition between individual and collective, work on affect attends to different forms of collective life (see Chapter 4).

11 These configurations go by different names: network, assemblage, apparatus, structure, meshwork, and so on. Some of these terms will be discussed at different points in the book, specifically section 2.3 and section 4.3.

12 Pile (2010) is right to claim a connection between work on emotions and work on affect around the question of relations. For Bondi (2005: 433), for example, we should approach emotion as 'as a relational, connective medium in which research, researchers and research subjects are necessarily immersed'. A proposition that coexists with slightly different assertions that we should 'theorize emotions relationally' (Bondi 2005: 434) or develop 'relational approaches to emotion' (Bondi 2005: 434). Anderson and Smith (2001: 9) attend to both 'emotional relations' and how 'social relations are lived through the emotions'. Thien (2005: 450) argues for 'placing emotion in the context of our always intersubjective relations', whilst Bondi, Davidson and Smith (2005: 3) 'argue for a non-objectifying view of emotions as relational flows, fluxes or currents, in-between people and places rather than 'things' or 'objects' to be studied or measured'.

and what they do. The initial task for an analysis of affective life is, then, to attend to differentiated 'capacities to affect and be affected'; exhaustion, pain, greed, and so on. The second task is to trace how affects emerge from and express specific relational configurations, whilst also themselves becoming elements within those formations. So, for example, greed is both an expression of particular relations and one part within the organisation of relations that we give the name capitalism to.

For an exemplification of this attention to relations, let us consider in more depth the greed of the capitalist that takes place in relation to, and determines, the exhaustion of the worker. Greed is the name Marx gives for an affect that follows from a specific relation with money. Money is, first, the independent and general form of wealth to be accumulated and, second, capital to be put into endless circulation (Casarino 2008). Casarino argues that greed for money takes two forms – miserliness and hedonism – that express the two relations with money in capitalism. Whilst clearly too straightforward, the example allows us to understand greed as expressing specific forms of relation and being an element in capitalism. It is not somehow a secondary illusion or a mere epiphenomenon. To paraphrase Massumi (2002a) on confidence, greed is as infrastructural to capitalism as a factory or a trading floor. As I suggested above, greed is both an effect of and an element within the value-producing activities given the name capitalism. But something that we might call greed also becomes akin to an affective condition; a collective mood that is imbricated with a repetition of the desire continually to accumulate capital that is integral to waves of capitalist construction and destruction.

This leads us to the difficult question of how to understand the relation between affect and other linked terms, such as emotion, mood or feeling. Is the greed of a capitalist the same as the greed of capital or of a capitalist society? Should we take them to be synonymous, in effect using terms such as emotion, feeling or affect interchangeably?[13] This is probably the most common use of the terms in human geography today, even though it means a lot of quite diverse experiences are gathered together in one shifting, ill-defined, category. The problem with this vague use is that it is too reliant on unspecified social and cultural assumptions about what specific terms mean and do. It is presumed that everyone will know exactly what the word 'emotion' is being used to name, for example. In contrast, should we insist on an ontological distinction between affect and other modalities?[14] The

13 The defence of this strategy would be that a difference between terms does not make a difference to research, thinking and politics. In fact, it does the opposite: leading to a forced distinction that obscures more than it reveals. This is a justifiable position. What is less justifiable is work that uses terms like emotion or affect vaguely and without either specification or justification.

14 In geography, this is the position that has been associated with non-representational theories, although in practice there is a range of positions within what is now a heterogeneous body of literature (see Anderson and Harrison 2010). There are also connections between an account of affect as bodily capacities and a range of what have been called 'new materialisms', including feminist rethinking of bodily substances (see Colls 2013).

terms affect and emotion have, in the main, fallen on one or the other of the divides between narrative/non-narrative and semiotic/asignifying (Ngai 2005). Affect with the impersonal, life and the objective; emotion with the personal, identity and the subjective. The turn to affect, on this understanding, is a turn to consider the non-conscious or not-yet-conscious dimensions of bodily experience through resources that cross between a hyperactive Deleuzian materialism and an experimentation with recent sciences of the body (Massumi 2002a; Connolly 2002).[15] However, beginning with an expansive definition of affect, as I have in the discussion of greed through Marx, unsettles an ontological distinction between modalities. Not because some form of distinction between affect and emotion cannot be made; every theory of affect and emotion makes some form of implicit or explicit distinction. Instead, it asks us to think about what sort of definition is offered and the work definitions do. Rather than an analytic distinction, it suggests we employ a 'pragmatic-contextual' (Ngai 2005) distinction that is designed to attend to the different types of experience gathered together in a unitary category such as 'affect'. This is to make terms such as feeling, mood, atmosphere, and so on, into sensitising devices designed to attend to and reveal specific types of relational configurations, rather than unproblematic claims about what affect really is 'out there' in the world or placeholders for an ontology of the personnel or impersonal.

Let us return to the example of greed to illustrate how a differentiated conceptual vocabulary might work to attune to different aspects of affective life. Perhaps, we could deploy the term *structure of feeling* to speculate on how an excessive greed seems to define the mood of a particular epoch in its relation to the accumulation and circulation of money (see Chapter 5). Perhaps, we could deploy the phrase *bodily capacities* to describe how greed becomes an embodied disposition that distorts and dominates an individual's relation to their life (see Chapter 4). The terms structure of feeling and bodily capacities are not used to make timeless claims about what affect really is. The question of 'what affect is' gets replaced by questions of what the terms allow us to do: What do they attune to? What do they show up? What do they sensitise thought and research to? A structure of feeling and a body's capacity are both relational, but as terms they sensitise us to different aspects of affective life. My hope is that a differentiated vocabulary enables us to attend to the complexity and multiplicity of affective life, whilst also opening up different ways of understanding how affective life is mediated, organised and occasionally surprises.

This pragmatic-contextual definition of affect – that begins from the question of what qualitatively different affects do – leads to a number of more formal starting points that emerge from my encounters with specific affect theories. In the main, these starting points have been associated with a range of non-representational theories. The name 'non-representational theories' – note the

15 The affective turn has been accompanied by work that has attended to a range of supposedly non-conscious modalities of experience, including sensation, perception, habit, memory and the senses (Paterson 2006; Wylie 2005; 2006).

plural – is used when gathering together a heterogeneous set of mostly, although not exclusively, materialist poststructuralist theories that share some loose starting problems, specifically: how sense and significance emerge from ongoing practical action; how, given the contingency of orders, practical action is organised; and how to attend to events and the chance of something different that they might open up (after Anderson and Harrison 2010: 23–4).[16] Cutting across these problematics, is a concern for the multiple ways in which affective life is made through processes of mediation. The term 'mediation' has generally been used in cultural and media studies to refer to something that stands in-between and reconciles two separate things. By comparison, non-representational theories consider mediation to be an ongoing process, occurring through multiple channels and forms (including affective conditions), and involving entities affecting one another in and through relations. This is to expand mediation beyond either the emphasis on the 'technical framing' of bodily capacities as we find in the work of Clough (2008), or the priming of affective dispositions through the media that is a shared focus of Massumi (2002a), Connolly (2002) and Thrift (2004a). These would be specific forms, effects and channels of mediation, in each case moving beyond a North Atlantic–Euro-modern equation of mediation with cognitive meaning (Latour 1993; Grossberg 2010). Mediation understood in this book is a general term for processes of relation that involve translation and change and from which affects as bodily capacities emerge as temporary stabilisations. In other words, mediation involves constant (dis)connections between affects and the complex mixtures that make up ways or forms of life. The processes through which affective life is mediated are many and varied, meaning that mediation as used in this book is not reducible to either a) mediation as the reconciliation of two opposing forces or b) mediation as equivalent to the media. The four starting points for an analytics of affect that follow from this account of mediation as a constant, active process are:

1. There is no such thing as affect 'itself' and no one affect (enchantment, hope, and so on) can serve as a model for an affect theory or affective politics. Instead, there are innumerable affects that are organised and patterned as part of diverse socio-spatial formations.
2. Affective life is always-already mediated; emergent from specific material arrangements that may be composed of all manner of bits and pieces. The geographies of affect will be a function of the relational configurations that object-targets, bodily capacities or affective conditions are both a part of and emerge from.
3. Affects such as an atmosphere of greed or an event of hope are not reducible to the material collectives that they emerge from. They have an efficacy

16 For summaries of non-representational theory/theories in geography, including the most significant critiques, engagements and developments, see Cadman (2009), Colls (2013), Thrift (2007) and Lorimer (2005; 2007; 2008).

as elements within those collectives. This means that affect is never autonomous. Affects are always-already imbricated with other dimensions of life without being reducible to other elements.

4. Representations are elements within those collectives. A representation may function as a 'small cog in an extra-textual practice' (Deleuze 1972 in Smith 1998: xvi). Attention to affect does not preclude an attention to representation and affect is not somehow the non-representational 'object' per se. Instead we must pay attention to how representations function affectively and how affective life is imbued with representations.[17] In addition, ideas about affect, or representations of affect, may take on an affective life of their own.

These four assertions are offered as initial provocations. I will elaborate on them and rework them throughout the book, before returning to them explicitly in the final chapter (section 7.2). They constitute not so much a theory of affect, as a flexible set of initial orientations for an analytics of affect that through an emphasis on mediation traces how affects 'can be, and are, attached to things, people, ideas, sensations, relations, activities, ambitions, institutions, and any number of other things, including other affects' (Sedgwick 2003: 19).

1.4 Organising and Mediating Affective Life

What implications do these starting points have for thinking through how affective life is organised? How do they disrupt or supplement other ways of thinking about the mediation of life and living, especially those connected to the emergence of affect as an explicit topic of concern in geography and elsewhere?

My emphasis on the organisation of affective life may seem odd. For what has been so promising about the turn to affect – where affect is understood as pre-individual intensities – has been the attunement to the unruly dynamics of living. From this work, we learn that new ways of living or moments of freedom are constantly appearing and being created amidst the 'to and fro' of everyday life. Whilst there are differences within theories that treat affect as pre-individual intensity, not least in how they think the relation between affect and signification or intention, they share a promise: attending to spaces of affect orientates inquiry to the real conditions under which new encounters, relations and events emerge

17 Deleuze goes on to distinguish himself from deconstruction (a method he nevertheless 'admires'), or at least a particular practice of deconstruction in relation to texts: 'It is not a question of commenting on the text by a method of deconstruction, or by a method of textual practice, or by other methods; it is a question of seeing what *use* it has in the extra textual practice that prolongs the text' (Deleuze 1972, translated in Smith 1998: xvi, italics in original). The comments resonate with work in geography and elsewhere that emphasises the practical work of making sense (Laurier 2010).

(see Massumi 2002c; Sedgwick 2003). The relation with life is in the main an affirmative one, bound up with the invention of new forms of witnessing able to sense and express the 'varied, surging capacities to affect and be affected that give everyday life the quality of a continual motion of relations, scenes, contingencies, and emergences' (Stewart 2007: 1–2; Swanton 2010; McCormack 2003; Holloway 2010).[18] The term affect orientates inquiry to the aleatory dynamics of an impersonal, unqualifiable, life that exceeds processes of mediation or organisation and is only imperfectly suggested in the names that we habitually give to psychological emotions or bodily feelings (hope, fear, and so on).

Obviously this link between affect, politics and contingency has engaged, interested and inspired me. As I have written elsewhere, I feel its political and ethical promise, even as I acknowledge that not everyone has or will (Anderson 2012: 29). For me, it offers a way of attending to moments of change in which social life is reordered and other possibilities may be glimpsed. Such an account matters because, as we shall see, it invites a particular style of engagement with the world: one that aims to sense and perhaps extend the potential for new ways of being and doing that events may open up. We see this attention to the excess of events across work that has attempted to bear witness to the potential for difference opened up by events that overflow various orderings; the fleeting potential that follows the event of a sexually-charged glance between two people (Lim 2007); the performative force and sense of mutability found in dance and the performing arts (Dewsbury 2000); the disruption of explicitly political events such as protests that attempt to break with the state of an existing situation (Dewsbury 2007; Woodward and Lea 2010). Conceptualised in this way, attending to affect opens up a series of questions about how to create, sustain and extend events (see Anderson and Harrison 2010). How can we relate to the potential that events perform and open up, the sense of promise and futurity that events may hold? How, put differently, can we relate to the future without capturing it and neutralising it before it happens? What affect-based research shares is a commitment to critique as a means of creating turning points in the here and now and a conviction that in any given situation more is needed than critique if those turning points are to be cultivated. Critique is necessary but always insufficient. It may be supplemented by a positive attachment to the existing world (see Woodyer and Geoghegan 2013). Hence the recent interest in enchantment (Bennett 2001) or generosity (Diprose 2001) as two such ethics of attachment to people, other animals and things. Whilst not always concerned with affect per se, we find a similar orientation to futures in the making in recent feminist and queer theorists' efforts to disclose potentiality within material life, and to develop inventive strategies for creating and caring for futures in the making (Muñoz 2009; Colls 2013).

---

18 Hence why the recent interest in affect has been accompanied by experimentation with new forms and practices of description attentive to the unforeseen and open to the world (see, for example, Swanton's (2010) use of montage in his work on race or Lim's (2010) use of vignettes).

Work on the politics of affect has in the main been concerned with one relation between affect and order: affect as excessive to the various ways in which life is calculated, specified, assessed, managed and otherwise intervened in. This involves thinking about the social as riven by chaotic, far from equilibrium, processes. Order is a precarious, hard-won, achievement, liable to break down at any point.[19] This much I agree with. It resonates with other currents in social and cultural thought. Nevertheless, there has only been limited attention to the processes whereby affective life is mediated and capacities to affect and be affected are organised. Even when claims are made that 'past actions and contexts are conserved and repeated' (Massumi 2002a: 30), we hear much more about affect's disruptive qualities. That is, how any organisation of affect is a subtraction from life's non-linear complexity that cannot but fail because of affect's excessive nature. Affect is presumed to be in excess of the ways in which life and living are organised and, as such, introduces a moment of indetermination into the mediation of life. So even if a body's 'affective charge' is constituted through the repetition of past contexts and actions, there is nevertheless always a 'slight surprise'[20] to affective life. For a body's affects are never fully determined, there is always an openness to them. Hence why Deleuze's affirmation, cited earlier, has been returned to and affirmed as a reminder of an ever present, never lost, reserve of potential in the here and now – a potentiality felt in the events that constitute a life.

The emphasis on the openness of affective life only takes us so far. Invoked in the abstract as a claim about 'affect itself', it risks being too general.[21] Worse than that, though, it risks offering an ahistorical and ageographical guarantee

19 Most notably, the emphasis on order as a hard-won achievement contingent on the holding together of diverse materialities connects some affect theories to the after-actor network theory literature on the coexistence of multiple realities (Hinchliffe 2007) and recent experimentations with the concept of assemblage that attempt to understand the relations between wholes and parts (McFarlane 2009).

20 The reference here is to Latour's (1999) phrase the 'slight surprise' of action. The point, as developed by Latour, clearly goes beyond a concern for affect specifically to the relation between the habits and practices of critique and how the social sciences might understand action outside of the dreams of mastery. He writes: 'I am always slightly surprised by what I do. That which acts through me is also surprised by what I do, by the chance to mutate, to change, and to bifurcate, the chance that I and the circumstances surrounding me offer to that which has been invited, recovered, welcomed (Julien 1995)' (Latour 1999: 281).

21 This guarantee has different grounds depending on the affect theory being deployed: it might be based on a claim about biology that selectively encounters and reads recent neuroscience (for example, Connolly 2002) or a claim about the virtual and the creative process whereby the virtual is actualised (for example, Massumi 2002a). Either way, both now intertwined approaches find a little too much hope that things might be otherwise and different and better in affect. I discuss the relation between affect and freedom in more detail in section 4.3 (see Anderson 2004b) for an early critique of the link between affect and excess based on an example of boredom).

that things might be different and better (Hemmings 2005). What is needed is an account of how affective life is organised and mediated that sits alongside the emphasis on the excess of affective life over and above existing determinations. This is the task of the book. It is complicated as soon as we attune to different understandings of affect. Returning to the three examples of hope, we find that affect is not one thing. If affect is simultaneously an object-target, bodily capacity and collective condition, then there may be multiple ways of ordering affect and multiple processes of mediation through which affects are imbricated with other processes.

This means exercising caution about affect theories that place the turn to affect in the context of claims about matter's capacities to emerge and self-organise (see Clough 2007). Recent work has drawn on accounts of non-organic life to consider the dynamism of affective life, rather than assume that affect is organised because bodies are pre-positioned within a signifying system. Applied to affect, the emphasis has been on describing affect as self-organising in the sense that form is generated immanently (Clough 2008). Whilst this approach has generated many insights, not least in questioning the emphasis on the body-as-organism, we must be cautious about assuming one general model for how affective life is mediated, organised and thereafter takes place. As we shall see, affect as object-target, bodily capacity and collective condition all involve qualitatively different forms and processes of organisation, albeit interconnected ones. For example, in relation to affect as an object-target we could ask: How are affects such as consumer confidence known through practices of knowledge? How are affects intervened in as subjects and collectives are rendered actionable and governed in specific power relations? In relation to affect as a collective condition we could ask: How do collective affects such as 'Obama hope' infuse, resonate with and become part of the institutions, practices and other things that make up society? How do collective affects condition social life, setting limits or exerting pressure on what is thinkable and doable? Finally, we might ask different questions about bodily capacities, principally how do 'capacities to affect and be affected' emerge, sediment or change in the midst of encounters and relations? And how do affective capacities involve representational practices such as speech acts or utterances?

Multiplying out the forms and processes of organisation learns from the insights of those non-representational theories that take order to be a problem and describe, carefully, step by step, how diverse human, non-human and other than human things are assembled in qualitatively distinct processes of mediation. If we begin from multiple partially connected translations of the term affect, then we find that we cannot base a theory of affect on one principle or form of mediation. On the contrary, we should trace multiple processes whereby affective life is mediated and understand how those processes, various as they may be, come into relation. In the following section I turn to describe how the book offers one approach to understanding affective life.

1.5 Problematics: Object-Target, Bodily Capacity, Collective Condition

Encountering Affect is organised around three translations of the term affect: affect as an *object-target* of apparatuses; affect as a *bodily capacity* emergent from encounters; and affect as a *collective condition* that mediates how life is lived and thought. These three translations do not add up to a new claim about the nature of 'affect itself', although they do develop from and elaborate on the starting points sketched out above. Nor is my aim to cover everything which now goes under the category of affective or emotional geographies. Rather, what the book attempts to do is outline some ways of thinking about the organisation of affective life, and offer concepts that might enable us to sense different processes of mediation. The argument about affective life builds through the book and each chapter is designed to discuss what up to that point has been implicit (although the vocabulary introduced in the later chapters is used throughout the earlier parts of the book).

The argument is exemplified and enlivened through examples of a range of affective geographies: the targeting of morale in aerial bombing; debility, dependency, dread in torture; hope in the midst of everyday life in an ordinary place in the UK; precariousness and insecurity in relation to changing organisations of capitalism and neo-liberalism; and emergency in how the UK state plans for events. What this purposefully diverse set of examples hold in common is that they all concern the presence of the future, albeit in different ways and with different implications. Futures become present as affective geographies are made and remade. By being made present a future, what has not and may never happen, comes to have real effects in the here and now. It is the same with affect. Seemingly vague, seemingly immaterial, affects such as hope, morale or precariousness achieve effects. And they do so across different domains of life: ranging from the violence of torture and aerial bombing to the ordinary goings-on as people listen to music. The relative weighting afforded the examples changes as the book progresses. In chapters 2–3, the presentation is designed to show how a theoretical vocabulary – derived principally from Foucault (2007; 2008) – allows us to trace how affective life is an object-target. Here, more emphasis is placed on the specificities of cases given that the theoretical vocabulary is designed to explicate the workings of apparatuses. Chapters 4–6 spend longer elaborating three concepts that make up a specific version of affective life – affect, structures of feeling and affective atmospheres. In these latter chapters, empirical detail is sacrificed in favour of theoretical development and elaboration. The examples across the book are deliberately diverse in order to emphasise that affect is not somehow the special property of one domain of life neatly separate from others. As well as presenting some of the difficulties in working with affect, the examples are discussed in a way that exemplifies different modes of engagement with life. Specifically, I show how description and speculation are necessary to attend to the dynamics of affective life, but should occur alongside a renewed practice of critique. Critique, description and speculation constitute a way of sensing, disclosing and perhaps intervening in the socio-spatial formations through which affect is mediated and ordered.

As a counter to the recent emphasis in some affect theory on immanent modes of (self)organisation, the first two chapters focus on how affective life is an object-target. Affects are known and become both objects and mediums for forms of intervention that aim to produce and reshape life. The example of consumer confidence and the tracking of optimism and pessimism about the economy is but one example of how affective life is taken to be a real force, rendered knowable through forms of calculation, and thereafter intervened in. In the case of the economic affect of 'consumer confidence', affect is rendered actionable at the level of the population understood as a collection of spenders and savers, for example. Chapters 2 and 3 offer a conceptual vocabulary for understanding this first mode of organisation and linked processes of mediation. Such an analysis is important because it avoids the temptation to see affective life as a kind of pure, unmediated, realm of experiential richness (Massumi 2002a) that would act as a counter, or bulwark, against the calculation and management of life. Instead, I argue that the task for inquiry is to map the 'intricate topology' (Toscano 2007: 120) whereby attempts to act on and through affect constantly bleed into and become part of affective life and vice versa.

Chapter 2 introduces Michel Foucault's concept of apparatus to understand how affects are 'correlated' with a range of non-discursive and discursive elements. Through a case study of the targeting of morale in a state of 'total war', I focus on how affect becomes an object-target for different forms of power. Rejecting an analysis which would counterpoise strategies for governing affect to the non-representational excess of affect, I show how forms of power can function through the indeterminacy of affective life. Beginning with the example of morale in a state of 'total war' is also designed to counteract the temptation to see the manipulation of affect as a defining characteristic of power today. The following chapter develops the Foucauldian argument that there is a dynamism to the relation between affective life and specific, diverse, modalities of power. Through an example of how the affective state of 'debility, dependency, dread' was named in early 1950s cybernetic behaviouralism before becoming central to interrogation and torture in the so-called 'War on Terror', I explore how specific 'versions' (Despret 2004) of affect and emotion are remade through apparatuses. Ways of naming, knowing, rendering actionable and intervening on affect are shown to be performative. Such an analysis is important as it invites us to specify how specific apparatuses function rather than making blanket statements that affective life is determined by power relations or general social categories. In the conclusion to Chapter 3 I suggest a set of questions for a practice of critique that would trace how apparatuses form and deform, emerge and change. Critique thereafter becomes a means of cultivating 'turning points', of bringing apparatuses to crisis. In contrast to recent work that has argued that critique has, in Latour's (2004a) terms, 'run out of steam', I argue that a specific practice of critique can sit alongside and complement speculation and description as ways of relating to affective life.

Chapter 4 is the bridge between the emphasis on affect as an *object-target* in chapters 2 and 3 and the emphasis on affect as a *collective condition* in chapters 5 and 6. I return to recent work in geography and elsewhere on spaces of affect and offer an analysis of affect as a *bodily capacity* emergent from encounters. Expanding the definition of affect as 'capacities to affect and be affected' introduced above, I show how affects form in the midst of the complex mixtures that make up specific encounters. Encounters include discursive and non-discursive elements that cannot be separated in practice. By way of moments of hope that first sparked my interest in theories of affect, I argue that the promise of attending to affect as a property of encounters is that it allows us to understand how ordinary affective life is ordered as it happens. This means that affect is not some kind of pure, unmediated, phenomenon, nor are affects transhistorical or transgeographical universals. Injecting a note of caution into approaches that would find a moment of politically-charged 'freedom' in affective life, I focus on some of the ways in which a body's 'force of existing' is mediated in encounters, without ever being fully determined. Apparatuses are one way in which encounters are organised, but only one. The affective conditions explored in chapters 5 and 6 are another way. However, Chapter 4 emphasises processes of mediation specific to how a body's 'force of existing' takes shape through encounters, including signifying apparatuses, the already existing habits and dispositions of bodies, and the presence and absence of other people and things. The chapter incorporates some examples of moments of hope that have long stayed with me and initially provoked me to engage with theories of affect.

Affect is not only an object-target or a bodily capacity emergent from encounters. Returning to the example of Obama-hope at the start of this chapter, perhaps hope was a mood shared within and beyond the site of the inauguration. An atmosphere emerges from a shared participation in the event whilst, simultaneously, atmospheres are part of the collective event of Obama's inauguration. Likewise greed may function as a collective affect. Excessive greed is frequently claimed as a characteristic emotion of capitalist societies as a whole, or of particular subsections within it. In these and other ways we are quite used to assuming that affects are collective phenomenon. We are also used to claiming that different types of collective affects can exist; think of how the mood of an era is identified across diverse spheres of life, for example. Chapters 5 and 6 explore how we might attend to the efficacy of collective affects and how affect as a collective condition might relate to other forms of organisation and processes of mediation.

Chapter 5 picks up Raymond Williams's (1977) notoriously ambiguous concept of 'structures of feeling'. Working it through a concern for the composition of collectives that marks non-representational theories and examples of precariousness amid neo-liberalising apparatuses, I describe how collective affects 'limit' and 'pressure' social life. Due emphasis is accorded to the processes through which collective affects are composed, change and coexist. Chapter 6 begins where the discussion of structures of feeling finishes: with a concern for the collective affects that emanate from specific sites and come to give those sites a particular 'enveloping' affective quality. Experimenting with the concept of 'affective atmospheres' via the work of phenomenologists Mikel Dufrenne and Gernot Böhme, I emphasise the ambiguous existence of collective affects. Affective atmospheres are the tension between presence and absence, materiality and immaterial, and the subjective and objective. Resonating with the insecurity of precariousness and the openness of the future found in moments of hope, the chapter is animated by an example of how the UK state prepares for disruptive events by staging and performing particular atmospheres of emergency. Chapter 6 ends with a discussion of how affects as collective conditions relate to affect as an object-target of apparatuses and affect as a bodily capacity emergent from encounters. I argue that collective affects both mediate apparatuses and encounters and are mediated by them, albeit in strange ways given the status of affective conditions such as structures of feeling and atmospheres as real and immaterial, present and absent, diffuse and structured.

What the conceptual resources I explore in this book share is a commitment to think about the organisation and mediation of affective life in a particular way. Turning back to the emphasis on non-representational theories, what they all share is a break with any assumption that life is organised through some type of transcendent form. By which I mean that emphasis is placed on how multiple processes of mediation work together, rather than finished orders. Following on, organisation happens through the holding together of diverse material and immaterial bits and pieces in distinct socio-spatial formations. It is only within this process of coming together and apart that the chance of difference is found and something new might be ushered into the world, however tentatively. The theories encountered in this book – Foucault's account of the operation of apparatuses of power, Sedgwick and Massumi's attention to the event of encounter, Williams's and various phenomenologists' sensitivity to collective affects as real conditions – share an attention to how mediation and organisation happen *immanently* as life takes place. Summarising this approach and emphasising how it enables an attunement to the variability of affective life, in the conclusion I set out how an analytics of affect might begin to understand how affective life is mediated. For to do so is also to ask how affective life is organised and to begin to invent and practice a mode of engagement appropriate to diverse forms and processes of mediation.

Chapter 2
Apparatuses

2.1 Object-Targets

In a symposium held in 1941 on the problem of how to maintain US national morale in the midst of war, an American philosopher, William Hocking, attempted to specify what morale is and does.[1] As we shall see, his attempt is but one of a number of efforts in the context of the Second World War to know the nature of morale in order to shape it. His answer is vague. Morale is 'something else' or 'something more':

> Morale, then, is something else than physical preparedness for an enterprise, something additional but not separable. … Morale itself, however, is something more than awareness of capacity, and a high morale may exist when capacity is low. Can we single out this something more? (Hocking 1941: 303).

Hocking's question – Can we single out this something more? – reverberated through the symposium as a group of then eminent social psychologists, military planners and urban theorists grappled with morale and its relation to war readiness and war fighting. It was deemed necessary to know, target and act on morale because of an urgent need: to hold together the nation – conceived of as a unified body – in the face of the uncertainties that 'total war' opens up. The threat was that panic would lead to a dissolution of the nation at a time in which it must act together. For even as morale was identified as a vague 'something more', it was nevertheless presumed to be central to life. The founder of the 1941 Committee for National Morale, Arthur Upham Pope,[2] put it in unequivocal terms: 'Morale wins

---

1 William Ernest Hocking (1873–1966) was an American philosopher principally interested in the philosophy of religion and associated with the idea of 'negative pragmatism'. He served as a professor of philosophy at Yale (1908–13) and then Harvard (1913–43), after studying with William James and Josiah Royce at Harvard. His book on morale (Hocking 1918) has its origins in his experience as a civil engineer in the First World War and his appointment as an inspector of 'war issues' courses in army training camps in northeastern USA. In *Morale and its Enemies*, Hocking (1918: IX) describes morale as a virtue – 'the practical virtue of the will to war' – and a kind of force – 'the invisible force behind war making'.

2 The Committee for National Morale was created in 1940 by art historian Arthur Upham Pope with the aim of founding a 'Federal Morale Service', to include a 'psychological research unit'. One of its first publications was a review of German psychological warfare (Farago 1941). Although such a service was never set up, not least because the army already

wars, solves crises, is an indispensable condition of a vigorous national life and equally essential to the maximum achievement of the individual' (Pope 1941: 195).

The attempt by Hocking, Pope and others to know morale is one example of how individual or collective affects become object-targets for action. There are many others in the context of a recognition by critics that affect and emotion are targets of intervention in late capitalist societies as a whole and within distinct sectors (see Hardt and Negri 2004; Hochschild 1983; Berlant 2011; Thrift 2007); friendliness is performed in service work (Hochschild 1983); satisfaction is tracked and measured through new forms of sentiment analysis; confidence or self-esteem are taught by private providers in workfare policies (Feher 2009); panic is predicted by specialists in crowd dynamics (Orr 2006); retailers attempt to modulate atmospheres (Miller 2012), to name but a few ways in which affective life becomes an object-target. In the next chapter I will describe how US and UK interrogation policies in the 'War on Terror' have attempted to create a specific affective state: 'debility, dependency, dread'. Understanding how affective life is known, rendered actionable and intervened in is therefore a pressing task for any analysis of affective life. Affective life is always-already 'informed' (Barry 2006) by attempts to know and understand it. Through the examples of morale and 'debility, dependency, dread', my aim in chapters 2 and 3 is to develop a conceptual vocabulary that can describe and critique how affects become object-targets. My emphasis is on the point where a relation of power meets a form of knowledge. By which I mean that various forms of knowledge are deployed to know affective life and various techniques are deployed to intervene in that life. This is, then, the first of the three ways of thinking about the mediation of affective life that I introduced in Chapter 1: affects are objects of knowledge, targets of intervention, and may be the means of intervening in life. With the result that affects – such as morale or 'debility, dependency, dread' – are 'inscribed in reality' (Foucault 2008: 20) as an effect of apparatuses rather than being 'things that exist, or errors, or illusions or ideologies' (20).[3]

Such a focus on affective life as an object-target might seem to jar with the attention to the indeterminacy of affective life that I argued in Chapter 1 has been so central to the promise of recent work on affect and emotion. For there is, on

---

had a morale branch, it occurred in the context of a huge interest in morale (particularly in relation to a unified 'national 'character'; see Bateson 1942) in the American Psychological Association and the American Association for Applied Psychology (see Watson 1942).

3 Foucault (2008) is discussing the 'existence' of the objects of the neo-liberalising apparatuses that he describes in *The Birth of Biopolitics* (such as homo economicus). He is also making a link to how his investigations of madness, disease, sexuality and delinquency 'show how the coupling of a set of practices and a regime of truth form an apparatus (*dispositif*) of knowledge-power that effectively marks out in reality that which does not exist and legitimately submits it to the division between true and false' (Foucault 2008: 19). Whilst the substantive context is obviously different, the methodological point concerning 'existence' still holds. I expand on it in this chapter and the next.

first reflection, nothing ambiguous about an object of power, nothing that resonates with the multiplicity, fluidity and openness that affect supposedly provokes cultural theory and cultural geography to think with and learn to sense. To describe how a named affect becomes power's object or an object of power is, on this account, to describe yet another way in which the openness of life is closed, reduced and contained in familiar processes of naming and classifying. Two qualifications are, therefore, necessary regarding my use of the term 'object of power' or 'power's object' throughout the book. First, an 'object of power' names the surface of contact for specific modes of power, and thus acts as a hinge between a desired outcome and the actions that make up the exercise of power. Yet, any exercise of power need not have an object in the sense of 'object' as the passive, reduced, effect of processes of abstraction, limitation and reduction. If we look at the etymology of the word object, we find a more unruly sense of object: object as an obstacle, something 'thrown in the way of', or 'standing in the way of' (Boulnois 2006). How an object of power shows up is, then, an open question. Second, establishing a surface of contact for power offers a solution to the problem of how to extend action into the future. As such, hopes, expectations and promises animate processes of knowing, naming and acting on an object of power. To return to the example of morale, targeting morale came to hold a promise and be invested with a hope: the promise of finding a better way of damaging or destroying an enemy and the hope that war can be won even in a state of 'total war', where war extends throughout life.

By way of Foucault's concept of apparatus, supplemented by his discussion in his late 1970s lectures on the 'existence' of the objects of apparatuses, I emphasise in this chapter and the next how object-targets come to be 'inscribed in reality' (Foucault 2008: 20). Through the case study of morale, my focus in this chapter is on how affective life is organised through multiple apparatuses that produce particular kinds of affective objects, particular kinds of subject and collectives, and are embedded in specific modalities of power. This will open up a set of questions about the performativity of representing and intervening in affective life that I take up in Chapter 3 through the example of 'debility, dependency, dread' and the 'War on Terror'. First, though, to some recent claims about the manipulation of affect by power today, in order to consider what kinds of things are intervened in when affective life is known, rendered actionable and shaped.

2.2 Manipulation and the Politics of 'Affect Itself'

The scent marketing company ScentAir UK sells the promise of atmospheric manipulation: the capacity to explicate the affective background of sites and turn atmospheres into resources to be harnessed for economic value-creation. Currently claiming upwards of 103 worldwide clients ranging from hotel chains to private hospitals, ScentAir manufactures 'scent systems and scent marketing solutions that are used by businesses to enhance the consumer experience'.[4] Various 'scent delivery systems' embedded in air conditioning systems deliver scents in tiny, ambient particles. Through over 1,500 scents such as 'earth', 'beach smells', 'pink bubblegum' the aim is the creation of atmospheres that act, to quote the company, 'just above the subconscious level of awareness'. As ScentAir describe themselves:

> Our scents enhance environments, identify brands and create memorable experiences. We do more than install air fresheners in lobbies, we work with each of our clients to establish and implement scent marketing strategies specific to their property.

Resonating with longstanding techniques of atmospheric modulation, think of the engineering of attention through retail atmospherics, ScentAir works to create intensities. 'Rock and Roll', 'coconut beach' and 'waffle cones' 'enhance guest experience' at the Hard Rock Hotel at Universal Colorado Resort, for example. 'Coconut beach' and 'Ocean' were used as part of the private medical company Celebration Health's transformation of its MRI imaging unit into 'Seaside Imagining': a beach-themed environment designed to remove the fear of MRI scans and thus cut cancellation rates, according to ScentAir's promotional material.

ScentAir could be interpreted as representative of a contemporary condition in which power now operates at the sub or just conscious level of bodily affects. The assertion is that subjects are now addressed as 'affective beings' (Isin 2004) and manipulated through new governmentalities that bypass consciousness. Slightly differently, but very often connected in practice, collectives of various types are addressed and acted on as 'affect structures' that feel as they act (after Isin 2004). States, institutions and corporations now know, target and work through affective life rendered actionable in one particular way: as bodily changes that although pre or non-conscious are nevertheless integral to the embodied dynamics of perceptual and sensory experience (Clough 2007, 2008; Thrift 2005). Contemporary power involves technologies that, as Clough (2008: 2) puts it, 'are making it possible to grasp and to manipulate the imperceptible dynamism of affect', specifically digital and molecular technologies allied to a resurgent neuroscience that is embedded in new forms of governance and value creation. ScentAir would – according to this argument – be an example of how techniques that know neurological or

4 All quotes in this section are from: www.scentair.com (last accessed 10 July 2012). ScentAir is the UK's biggest distributor of artificial smells as part of the burgeoning field of 'scent marketing': the strategic use of scent and olfactory experience in relation to commercial products. Linked to changes in technologies for the production and distribution of scents, scent is used to shape consumer action, specifically attention (as measured through 'dwell time' in a 'retail environment') and acts of purchase. The use of scent is part of a recent recognition that branding needs to work through the emotions (with scent marketing itself marketed through the promise that smell works directly on the limbic system), but also forms part of a longer term emphasis on the active shaping of retail atmospherics.

biochemical processes are being deployed in the commercial sphere as part of a nascent attention economy, where attention is simultaneously a scarce commodity, a valuable resource and a form of capital (Terranova 2012). The argument follows that there is now a pressing need to understand how power works affectively – that is in ways that are not equivalent to ideology (at least if ideology is understood as the mediation of cognitive meaning).[5]

These are important claims. Ones that place attempts to intervene in and through 'affective beings' or 'affect structures' in the context of, first, transformations in biopolitics associated with neo-liberal modes of governance and, second, the invention of new sources of profit in today's post-Fordist capitalism.[6] Some critics have gone further to identify new, emergent political formations. Lazzarato (2006: 186), for example, connects a new politics to transformations in value: 'Noo-politics (the ensemble of techniques of control) is exercised on the brain. It involves above all attention, and is aimed at the control of memory and its virtual power'. Underpinning these claims about power and affect today is an assumption that 'life itself' is now understood in terms of self-organisation, morphogenesis and recombination, rather than homoeostasis and equilibrium (see Dillon and Reid 2009). The thesis is that value is generated, and life governed, through what Clough (2008) terms an information-imbued 'biomediated body': a body that may be produced, managed and experimented with through techniques ranging from new media and information technologies to affective psychopharmacologies (Parisi and Goodman 2011).[7] As such, this work on power and 'affect itself'

---

5 Massumi (2002a: 263) understands ideology in the following specific sense, one that downplays the passions of ideology: 'Ideology is construed here in both the commonsense meaning as a structure of belief, and in the cultural-theoretical sense of an interpellative subject positioning'.

6 Whilst not my focus here, it is worth noting that various diagnostic terms have been used to make sense of the relation between affect as an object-target of manipulation and transformations in post-Fordist capitalist value-creation and profit generation. These include ways of categorising the dominant or hegemonic structuring of value producing activities – 'attention economy', 'affective economy', 'immaterial production' (Terranova 2012) – the identification of new types of work and labour – 'immaterial labour', 'affective or emotional labour' (Hardt and Negri 2004) – and the naming of new modalities of power – 'preemptive power' or 'mnemonic control' (Parisi and Goodman 2011).

7 Here we should note two intertwined trajectories of work on affect/emotion and labour, both of which centre on questions of affect and value in the context of work and changes to capitalism. First, work on forms of emotional labour in a post-industrial economy, including the marketisation of care and other intimate relations (Hochschild 1983; 2012). Second, the tradition of Italian feminist work on the intricate connections between material and immaterial labour in the context of debates about the changing production of value in post-Fordism (Fortunati 1995; 2007). Contra to the distinction between affective and other immaterial labour offered by Hardt and Negri (2001), this work shows the problems of categorising forms of labour according to whether the products are 'material' or 'immaterial' (see Fortunati (2007) for a critique of the distinction between immaterial

complements longstanding feminist concerns with how power works through emotional life by diagnosing how new forms of biopolitical control measure life capacities, manipulate molecular life at the level of affect and result in divisions between valued and devalued lives.

The recent interest in affect within the social sciences and humanities could be seen as both a response to, and symptomatic of, these changes. They have certainly provided one key rationale for learning to attend to affective life. Massumi (2002c: 233) states directly that 'Affect is now much more important for understanding power, even state power narrowly defined, than concepts like ideology. Direct affect modulation takes the place of old-style ideology'. Thrift (2004a: 54) similarly highlights how such affective manipulation occurs across the domains of politics and economy, arguing that 'the discovery of new means of practicing affect is also the discovery of a whole new means of manipulation by the powerful'. Clough (2007: 2) is more specific than Massumi or Thrift by arguing that thinking in and through affect is necessary to grasp '[t]he changing global processes of accumulating capital and employing labor power through the deployment of technoscience to reach beyond the limitations of the human body, or what is called "life itself"'.

The strength and scope of these claims about the relation between power and affect give me cause to pause, even as I am inspired by how they tie affect into a searing critique of the transformations of power in the contemporary condition. One problem is that claims that individuals are being governed affectively are not new given the relation between neo-liberal or advanced liberal governmentalities (including values of freedom and autonomy) and the psy-disciplines (Rose 1998). Neither is the concern with the affective manipulation of subjects novel. For example, in 1957 Vance Packard warned of what could then be considered a 'rather exotic' new 'motivational research' industry that manipulated 'inner thoughts, fears and dreams' at subconscious levels where 'we not only are not aware of our true attitudes and feelings but would not discuss them if we could' (Packard 1957: 28).[8] A more pressing issue, as Barnett (2008: 190) has rightly identified, is that (some) of the work on power and affect has reproduced a 'recurrent trope of

---

labour based on the intellect and cognition and affective labour based on the body). For example: 'Showing one's affection for a person means possibly setting off on a journey, buying a present or preparing a dinner, it means following an immaterial expression with concrete acts' (Lorente 2004 in Fortunati 2007: 140).

8 To an American public who at that time were animated by concerns over Communist brain-washing, the techniques of the 'hidden persuaders' were dammed as immoral, even a threat to Cold War America because they removed from people their 'right to decide … what they want to do and who they want to be' (Packard 1957: 28). Packard's persuaded subject was moved by 'triggers of action' (27) (such as words, acts or phrases) without deliberation. She (because Packard's moved subject is a gendered one) was susceptible to a veritable army of 'hidden manipulators' who addressed subjects as affective beings composed of '[b]undles of day-dreams, misty hidden yearnings, guilt complexes, irrational emotional blockages' (14).

manipulation', where it is presumed that manipulation is undertaken strategically and instrumentally by coherent, but vaguely specified actors (including the media). The result is a one-dimensional account of the relation between power and affective life. To avoid assuming action on and through affect is equivalent to manipulation, we should multiply out the modalities of action whilst, at the same time, taking care to avoid presuming the success of those modalities of action in shaping affect. To this end, I would sound two further notes of caution about any claim that the manipulation of affect is new and somehow unique to the contemporary condition.

First, power has not somehow just discovered affective life. On the contrary, the entrance of life into power's orbit involves action on and through affective life: from the eruptive passions of the mad moulded by the power of normalisation to the aleatory movement of public opinion regulated at the level of collectives, power works affectively, albeit in various, distinct, ways. For one example, consider Foucault's (1977) discussion in *Discipline and Punish* of the relation between the ceremony of punishment, in particular public tortures and executions, and sovereign power. Foucault writes of public executions and tortures as a 'policy of terror' (49) that affirmed the 'dissymmetry of forces' and worked through the activation of public feeling amongst those called to witness and participate in the spectacle:

> In the ceremonies of the public execution, the main character was the people, whose real and immediate presence was required for the performance. An execution that was known to be taking place, but which did so in secret, would scarcely have had any meaning. The aim was to make an example, not only by making people aware that the slightest offence was likely to be punished, but by arousing feelings of terror by the spectacle of power letting its anger fall upon the guilty person (Foucault 1977: 58).

Compare with the form of punitive power that replaces punishment-as-spectacle, even if a 'trace of torture' (Foucault 1977: 16) remains in modern criminal justice.[9] Punishment's affective grip changes as it loses its glory: 'its effectiveness is seen as resulting from its inevitability, not from its visible intensity; it is the certainty

9 In hints to the complex affective geographies of state power, Foucault stresses that the public executions and tortures were also sites for a rejection of punitive power and occasionally revolt (58–65). The 'scene of terror' became something else, a site for state power to fear; 'In these executions, which ought to show only the terrorizing power of the prince, there was a whole aspect of the carnival, in which rules were inverted, authority mocked and criminals transformed into heroes. The shame was turned round; the courage, like the tears and cries of the condemned, caused offence only to the law' (61 – earlier Foucault (8) describes public torture/execution as 'a gloomy spectacle of punishment'). Foucault goes onto show how, contrary to the designs of state sovereign power, glory and abomination come to coexist around the criminal through the practice of 'gallows speeches', in which the condemned person made a proclamation of his/her crimes (65–9).

of being punished and not the horrifying spectacle of public punishment that must discourage crime; the exemplary mechanics of punishment changes its mechanisms' (9).[10]

Second, and following the emphasis on the variegated types of power, we should also be cautious about claims that the manipulation of subjects as 'affective beings' or collectives as 'affect structures' is a single or coherent phenomenon that can be easily mapped onto contemporary political-economic transformations. Too often, critics claim that in a supposedly neo-liberal age governmentalities bypass intentionality to shape directly the 'plasticity' of the body-brain nexus. Undoubtedly, this is the case for some techniques that have begun to draw on the neurosciences. Jones, Pykett and Whitehead (2011), for example, carefully trace new forms of psychological governance that align neuroscience and behaviouralism and (re)problematise intractable social problems as pathologies of behaviour. Nevertheless, we have to be specific about how affective life is known and intervened in, how other kinds of objects are produced in that process, and how exactly ways of knowing and intervening in affective life involve distinctive, albeit partially connected, modalities of power.

Let us return to the example of ScentAir, and avoid reducing it to an example of a set of political-economic transformations that govern through molecular life. In order to shape a site's atmosphere, affective life is housed in two referents connected by the olfactory system – environments and the neurological body. Specific forms of knowledge are used to know affective life (neuroscience) and varied techniques are deployed to intervene in environments and bodies: branding strategies that give a quality to products, sites or people and the introduction of scent as a new atmospheric materiality that (re)shapes the affective background. Compare with the example of 1950s advertising that is the object of concern in *The Hidden Persuaders* (Packard 1957). What Packard deplores is the use of a type of practical psychoanalysis that renders the subject actionable by purporting to know the unconscious. What these brief examples give a sense of is not only the range of knowledges of affective life that exist and are drawn upon, but also the way in which they render life actionable – subject to some form of intervention – in varied, only partially connected ways.[11] And that the forms of intervention that

10 Foucault (1977: 9) writes of the affective relations that punishment is embedded within. For example, writing of spectacle and sovereign power: 'in punishment-as-spectacle a confused horror spread from the scaffold; it enveloped both executioner and condemned; and, although it was already ready to invert the shame inflicted on the victim into pity or glory, it often turned the legal violence of the executioner into shame'.

11 For another example, compare with the still popular idea of 'emotional intelligence'. 'Emotional intelligence' interweaves a diagnosis of societal-wide emotional malaise and breakdown (including depression and aggression) with an emphasis on the changeable character of what Daniel Goleman (1996) terms 'underlying emotional competences'. For Goleman (xii), these centre on 'self-control, zeal and persistence, and the ability to motivate oneself'). Whilst the idea of 'emotional intelligence' draws on 'new discoveries about the

follow from this process – whether through practical psychotherapies or atmospheric modulation – cannot be reduced to a broad trope of the manipulation of subjects shorn of their ability to deliberate and decide. Manipulation is one modality of action that takes place alongside others. ScentAir involves a different modality of action: a kind of environmental design. Rather than entraining bodily capacities or regulating populations, the milieu of action is the object and target for an intervention. Following Ash (2012), we could say that ScentAir works by setting up possibilities for action within a set of thresholds. It acts indirectly on subjects by shaping the affective quality of a 'retail environment'. Compare with the 'visible intensity' and affirmation of sovereign power that Foucault (1977: 9) finds in sites and scenes of public execution and torture.

In summary, I want to remain open about the relation between power and affect and avoid reducing that relation to a single story about today's supposedly 'affective society'. We reach, then, a key question once we start from the multiplicity of relations: how is affective life constituted as a specific kind of object-target and with what effects? In the following section I turn to the thought and work of Foucault for a method that might enable us to answer this question without recourse to an epochal account of the contemporary in which it is presumed that manipulation of affect is a new phenomenon, or it is claimed that power now works through molecular life and bypasses intentionality. In the recently translated *Security, Territory, Population* lecture series (2007: 239) Foucault describes his method as one which involves 'showing' the 'coagulation, support, reciprocal reinforcement, cohesion and integration' of a differential field made up of a 'multiplicity of extraordinarily diverse processes' (238). It is this emphasis on showing the gathering of a multiplicity that is in the background to the examples of morale and 'dread, debility and dependence' in this chapter and the next. First, though, to Foucault's concept of apparatus and how it enables us to map attempts to mobilise, impede, shape, induce, stop or otherwise act on affective life.

2.3 Apparatuses

To turn to Foucault for a method might appear to be slightly against the grain, given the general disavowal of him by some affect theorists (see Sedgwick 2003) and the recent charge that he had a 'seeming aversion to discussing affect *explicitly*' (Thrift 2006: 54, emphasis added). Nevertheless, something like affect reappears as a concern throughout Foucault's archaeologies and genealogies, albeit perhaps an implicit one.[12] One of those occurrences is, I think, of especial significance for how it hints towards a method for tracing how a multiplicity of differential elements

brain's emotional architecture' (xii), it also owes much to the self-help genre as a technique of the self that encourages work on the emotions.

12 For one example, consider Foucault's (2006) passage on The Great Fear from the re-translated *History of Madness* that hints to a relation between collective affects and the

come together and hold together. In the lecture series from 1978–79, *The Birth of Biopolitics*, Foucault (2008) ties liberalism and security to the production of a kind of generalised, ambient, fear. The details of his argument concerning liberalism need not concern us here. What is of interest is Foucault's method. He describes this 'fear of danger' in the following terms in the lecture of 24 January 1979:

> that is to say, individuals are constantly exposed to danger, or rather, they are conditioned to experience their situation, their life, their present, and their future as containing danger. ... In short, everywhere you see this stimulation of the fear of danger, which is, as it were, the condition, the internal psychological and cultural correlate of liberalism (66–7).

As with so many other intriguing ideas in the lectures, the 'stimulation of the fear of danger' is only touched on. We hear nothing about the specific mechanisms through which 'fear of danger' is stimulated (or 'conditioned', to use the other phrase in the above passage). Nevertheless, Foucault's brief reflections invite us to think seriously about how the 'stimulation of the fear of danger' – and other ways of acting on and through affect – are linked to specific processes of mediation. He appears to afford something like 'fear of danger' a double status in this passage and the surrounding lectures. Fear is simultaneously an effect of an apparatus and one component amongst others within an apparatus. As such, it provides the 'correlate' at the level of the 'psychological and cultural', admittedly a vague formulation, for the emergence, deployment and intensification of liberal apparatuses of security.

Foucault's claim about 'fear of danger' might appear, on first reading, to chime with various attempts to diagnose the contemporary in terms of a characteristic collective mood or set of closely linked collective moods (this work is discussed in section 5.2). Foucault's brief comments do lend themselves to such an epochal understanding of the relation between collective affect and the contemporary. Nevertheless, I begin with them here because they also prompt us to offer a non-epochal account of the contingent relations between affective life and the different ways in which, to quote Agamben (2009: 12), attempts are made to 'manage, govern, control, and orient – in a way that purports to be useful – the behaviours, gestures, and thoughts of human beings'. This will involve, however, taking Foucault beyond Foucault by expanding on one term in the above discussion – 'correlate'– and linking it to the concept of apparatus. 'Fear of danger' is described

---

conditions of possibility for complex transformations in the emergence of the mad and the treatment of madness. To cite just one passage:

> Suddenly, in the space of a few years in the mid-eighteenth century, *a fear emerged*. It was a fear formulated in medical terms, but deep down it was animated by a whole moral mythology. People were in dread of a mysterious sickness that apparently emanated from houses of confinement and was soon to spread throughout the cities. (355, emphasis added)

by Foucault as the 'correlate' of liberalism. Intriguingly, we find him (2007) using the term 'correlation' in the previous set of lectures in a discussion of method (see Collier (2009) for a systematic discussion). Instead of an epochal account that would trace the replacement of a 'disciplinary society' with a 'society of security', Foucault advocates a focus on the 'system of correlations' between diverse elements. What he emphasises is less a categorical break with other modes of power, or some simple relation of succession, and more *redeployments* and *intensifications* of elements from other modes of power together with the *emergence* of new ways of securing life (Nealon 2008). Thus:

> There is not the legal age, the disciplinary age, and then the age of security. ... In reality you have a series of complex edifices in which, of course, the techniques themselves change and are perfected, or anyway become more complicated, but in which what above all changes is the dominant characteristic, or more exactly, the system of correlation between juridico-legal mechanisms, disciplinary mechanisms, and mechanisms of security (Foucault 2007: 8).

Following Foucault's attention to such 'systems of correlation', my starting point is that affective life becomes an object-target as part of the 'complex edifices' that are apparatuses rather than as part of some kind of epochal change (such as the birth of an 'affective society' intimate with new knowledges of 'life itself'). In this context we can better understand Foucault's use of the term 'correlate' in the lecture of 24 January 1979. Object-targets – in the case Foucault discusses 'fear of danger' – are 'correlated' in the sense of being 'inscribed within reality' in relation to the institutions, materialities, techniques, people and much more that compose an apparatus. It is from within apparatuses that affective life becomes an object-target. Apparatuses produce specific versions of what affect, emotion and other modalities are and do (and I will say more about 'versions' (Despret 2004) in the next chapter). Starting from apparatuses requires a displacement of analysis from a specific actor who manipulates affective life to the ongoing integration of a multiplicity of elements, the emergence of specific functions and the constitution of intentions.

Foucault (1980) offers the closest he gets to a definition of 'apparatus' in a round-table interview with historians conducted in 1977.[13] It is worth quoting Foucault at length, as he draws out three specific elements of the term that I want to pick up on and elaborate. Together, they enable us to differentiate 'apparatus' from other linked terms such as assemblage, network or mesh which similarly emphasise how life is mediated in relation:

---

13 As Elden (2001) shows, the french *dispositif* resists translation, or rather has provoked various attempts to translate it: 'construct', 'deployment', 'grid of intelligibility', 'device' and 'social apparatus' have all been used.

> What I'm trying to pick out with this term is, firstly, a thoroughly heterogeneous ensemble consisting of discourses, institutions, architectural forms, regulatory discourses, laws, administrative measures, scientific statements, philosophical, moral and philanthropic propositions – in short, the said as much as the unsaid. Such are the elements of the apparatus. The apparatus itself is the system of relations that are established between these elements. Secondly, what I'm trying to identify in this apparatus is precisely the nature of the connection that can exist between these heterogeneous elements. Thus, a particular discourse can figure at one time as the programme of an institution, and at another it can function as a means of justifying or masking a practice which itself remains silent, or as a secondary re-interpretation of this practice, opening out for it a new field of rationality. In short, between these elements, whether discursive or non-discursive, there is a sort of interplay of shifts of position and modifications of function which has as its major function at a given historical moment of responding to an *urgent need*. The apparatus thus has a dominant strategic function (Foucault 1980: 194–5, emphasis in original).

The first point to draw out from Foucault's response is that an apparatus is materially heterogeneous, composed of a set of discursive and non-discursive elements. There are no a priori limits to the type of entities or relations that are to be found within an apparatus. In the 1977 interview, Foucault (1980: 197) clarifies that the 'episteme' is a specifically 'discursive' apparatus based on diverse kinds of statements or utterances, while the apparatus in its 'general form' includes both discursive and non-discursive elements. This gives us a first methodological orientation, albeit now a fairly familiar one: affective life emerges as an object-target through and alongside a heterogeneous array of elements. If we return to the example of ScentAir then we can get a sense of this material heterogeneity. Included within the apparatus would be: the phenomenon of branding, the scent delivery system, concepts of free will, a link to Walt Disney, the international perfume houses, the figure of the consumer, particles, and so on. For 'consumer neuroscience' more widely it would include the electrodes attached to heads, the promise attached to neuroscience, cuffs on fingertips, printouts of brain scans, and so on.[14] Importantly, the differential elements would extend beyond those that occupy a particular site. They might include a whole series of calculative devices for knowing the unknown consumer, ideas about the sovereignty of the consumer, the organisation of a consumer economy, and a series of experimental and theoretical knowledges based on new brain sciences, amongst much else.

The emphasis on material heterogeneity is now common to various types of relational thinking and would make the term apparatus equivalent to linked terms that also attempt to capture a sense of ontological diversity; mesh, assemblage, network, for example. Like other forms of relational thought, Foucault lays

14 On the promise and rhetorical power of brain imaging in the context of what the authors term 'neuromania' see Legrenzi and Umilta (2011) and Thornton (2011).

emphasis on how an apparatus will generate intentional or unintentional 'effects' that enter into 'resonance' or 'contradiction' with other effects. This means that apparatuses have a life: the 'elements' that make up a differential field are constantly being redeployed, readjusted, reworked or intensified. To be more precise, though, Foucault (1980) describes the apparatus as the 'network' between these discursive and non-discursive elements. We can take the term 'network' to be equivalent to the term 'system of correlation' used in the lectures series (Foucault 2007: 8). Whilst perhaps giving too strong a sense of a system of nodes and lines, the term 'network' is best thought of as a characteristic relation or set of relations through which the differential elements that make up an apparatus come together and may be held together. In other words, an apparatus involves what Bennett (2010) terms a 'style of structuration': a distinctive connection between a set of differential elements that integrates those elements through the combination of 'strategies of relations of forces supporting, and supported by, types of knowledge' (Foucault 1980: 196). As such, use of the term apparatus gives more of a sense of the ongoing *integration* of a differential field of multiple elements than linked descriptors such as assemblage (see De Landa 2005; McFarlane 2009; Legg 2010).

The integration of differential elements takes a specific form. Apparatuses initially derive from a 'strategic function' or 'strategic objective' that responds to a specific 'demand'. Apparatuses 'continue in existence' through a perpetual process of 'strategic re-elaboration' or 'strategic completion' (Foucault 1980: 195). Unintended or negative effects will be enrolled into a new strategy (referred to by Foucault as 'strategic elaboration'). Let us return to the examples of ScentAir to illustrate this idea. At the heart of the deployment of scent as a new atmospheric materiality in branding strategies is the strategic problem of the unknown consumer and the ever present demand to 'get closer' to the consumer. Whilst figured as a subject who can be manipulated, the consumer is simultaneously a subject who evades attempts to know him or her. In part this is because he/she does not know their own motivations. In part it is because if they did know their own motivations they may not reveal them to the developers of products, services and experiences. It is to enrol the consumer in a relation of purchase that the background affective quality of sites are explicated and shaped and affective life is intervened in through retail atmospherics, including but not limited to scent (with no guarantee of correspondence between intention and effect).

There are some important issues to be picked up and developed later on in the book; not least whether the dynamism of affective life is best understood in terms of the interconnected processes of strategic re-elaboration on the one hand and the creation of intended and unintended effects on the other. Leaving this question for the moment, the concept of apparatus provides an important starting point for attempts to understand how affective life is mediated and produced as an object-target. It orientates inquiry in four specific ways. First, to the specific 'demand' or 'urgent need' that the apparatus responds to and emerges from. This 'demand' or 'urgent need' is outside of the apparatus, can be thought to pre-exist it, but becomes causal in the initial constitution of an apparatus (it also may change as an apparatus 'continues in existence'). Second, to the strategic 'function(s)' or 'objective(s)' in relation to which a differential field is integrated. Third, to the heterogeneous elements and relation(s) between those elements that comprise the apparatus. Fourth, to the constant process of strategic (re)elaboration that happens as intended and unintended effects are generated, come into relation with one another, and lead to (re)deployments or intensifications.

These are only general methodological starting points. What they offer – or, perhaps better, promise – is a first step in thinking what I described in Chapter 1 as the mediation and the organisation of affective life. My claim is that part of those ways of mediating life are multiple, partially connected, apparatuses that attempt to structure the capacity to affect and be affected of people and groups. Apparatuses might include the great (de)individualising apparatuses of discipline described by Foucault (1977) – the school, factory, army – or the new apparatuses of biopolitical control that now extend beyond enclosed spaces in fluctuating, never ending, modulations. For example, Stiegler (2010: 126) writes of the specifics of a 'psychotechnological psychopower' based on 'establishing markets for consumption' (128). Ash (2012), drawing inventively on Stiegler, shows how the modulation of affect is central to the commercial success of the apparatuses of video game production and consumption. Developing an analysis of the 'retentional economy' of video games, Ash argues that designers utilise technical systems and environments in an attempt to transmit and translate potentials for affect. Juxtaposing Ash's Stiegler-inspired analysis of a psychopower oriented to attention as a resource for capital with the more well-known disciplinary apparatuses reminds us of the multiplicity of apparatuses and their coexistence. What kind of object-target is affect as constituted through apparatuses and how do apparatuses constitute other kinds of objects alongside named affects? I return to the example of morale and aerial bombing to address the question. For now, though, perhaps we can think of the object-targets of apparatuses as having the same status of 'existence' that Foucault (2008) carefully, if a little awkwardly, gives the objects of the neo-liberalising apparatuses he describes in *The Birth of Biopolitics* lecture series. He explains this in a methodological note in the lecture given on 10 January 1979. It is worth quoting it at length to place the curious formulation Foucault ends with in the context of the operation of apparatuses:

> The question here is the same as the question I addressed with regard to madness, disease, delinquency, and sexuality. In all of these cases, it was not a question of showing how these objects were for a long time hidden before finally being discovered, nor of showing how all of these objects are only wicked illusions or ideological products to be dispelled in the [light] of reason finally having reached its zenith. It was a matter of showing by what conjunctions a whole set of practices – from the moment they become coordinated with a regime of truth – was able to make what does not exist (madness, disease, delinquency, sexuality, etcetera), nonetheless become something, something however that continues not to exist (Foucault 2008: 19).

Perhaps morale, dread, debility, dependency and other object-targets also become 'something, something however that continues not to exist' (19)? For me, Foucault's formulation is useful because it avoids making affect as named, known and acted on into either an object of the natural sciences or an object of social constructionism. Instead, affect 'exists' in the sense that affects are 'inscribed within reality' (19) through apparatuses that know, render actionable and intervene in life. For example, affect is not the same thing and does not have the same effects in all examples of the commercial deployment of consumer neuroscience more widely. For ScentAir, affects are at the threshold of consciousness and linked to olfactory entrainment and the nervous system. For other examples of neuro-marketing, affects are below the threshold of consciousness but linked to neurological patterns (Thornton 2011). What is constituted are various, albeit partially connected, accounts of what affect is in relation to the same strategic function of knowing the unknown consumer and in relation to the urgent need of capital accumulation. We must therefore begin analysis with a declaration of ignorance: we do not know how affects and other kinds of objects are constituted outside of specific apparatuses. The following section turns to illustrate this approach and, implicitly, recognise some of the difficulties of analysis by following how morale was rendered actionable as a particular kind of object-target through the apparatuses of 'total war'. The example of morale in 'total war' is also designed to challenge the presentism in analyses that understand affective manipulation as symptomatic of biopower today, and remind us that claims about the supposedly non-representational, excessive nature of affect and affective life are far from new.

2.4 Morale in a State of Total War

In the context of the emergence of 'total war', governments invented ways of targeting and destroying morale and ways of protecting and harnessing it. It is in this context that specific apparatuses come to act on and through morale. The term 'total war' was first popularised by Erich Ludendorff in a pamphlet *Der totale Krieg* during the First World War. But its first use was by French civilian leaders in the Great War who coined the terms *guerre totale* and *guerre integrale* (Chickering, Förster and Greiner 2004). 'Total war', as used to designate a concrete historical phenomenon rather than ideal type of war, involved two changes that made war 'a war of nerves' that altered the 'character of peace' (Park 1941: 360). First, there is an expansion of the front line of war through the advent of new extended technologies of destruction and damage that reduce the distance between home front and front line, folding the two into one another. Strategic bombing, for example, involved an asymmetry between the destructive capacities of the bomber and the vulnerability of the bombed. It opened up the possibility of occupation by air (Lindqvist 2002). Psychological operations, slightly differently, involved techniques such as rumour or misinformation that aimed to shape and mould perception. These worked through and subverted the various channels of information that made up daily life (such as radio, posters, newspapers, and so forth.) (Virilio and Lotringer 1997). Second, distinctions between civilian and soldier, combatant and non-combatant, tended to fade or be eliminated as mobilisation for war was 'total'. War extended throughout the spaces of the economy or leisure and, consequently, came to rest on the participation of populations. The battlefield is extended. A 'home front' is established that is made up of new actors such as 'industrial workers' or 'domestic workers'. These are variously protected (through the architecture of shelters, for example) and targeted (in aerial bombing or by rumour generation). These two changes make war 'total' in the sense that the apparatuses of the state aim to expand to every sphere of life and all of life must, consequently, be mobilised for the war effort (Van Creveld 1991).

A set of affects accompanied these changes in the spatial form of war, of which morale is but one part. These include the pleasures and passions of the destructive activities of 'total war' and the various attempts to regulate those passions through ideals such as 'honour' or 'glory' or through disciplinary practices such as the drill (Ehrenreich 1997; Burke 1999). They also extend to the traumatic experiential geographies of suffering or loss that can haunt the victims and sometimes perpetrators of the multiple relations and forms of violence that make up 'total war' (Hewitt 1994). Whilst normally considered to be a pacifist manoeuvre – as it makes present the horrors of war – it is also worth noting that understanding the 'total' battlefield as a site of swirling, resonating, affects has been central to fanatical praise of 'total war'. 'Total war' becomes an event that reveals the inhuman forces that undo and disperse the fragile form of the human (Toscano 2007). Ernst Junger's call for 'total mobilisation' in his fascist 1920 *Storm of Steel,* for example, finds in war's sundering of the comforts and habits of individuality what Toscano (2007: 189) critically terms an 'intensity-in-movement devoid of any intrinsic organic armature, a vitalism that only appears at the very limits of organism, whether this be physiological, political, or aesthetic in nature'.

The spaces of 'total war', from the battlefield to the home or trenches, are spaces of intense affect and this has long been recognised in forms of military thinking. Morale, though, is unique: both because it is born in the emergence of a new dimension of war – 'intense fellow feeling' as part of warfare (De Landa 1991) – and because it has subsequently accumulated a promise in Western military thinking. Morale is the indeterminate and indefinable target that, whether destroyed or protected, would enable the activity of war to carry on the momentum of its own enforcement. A certain anticipatory tone has long infused and animated attempts to render morale actionable: a sense of possibility that we could name as hopefulness. The following account traces one episode in the history of this promise, describing how morale emerged as a diffuse potentiality to be secured as part of the excess of providential and catastrophic institutions and operations that make up 'total war'. Civil protection and aerial bombing emerge in response to a shared 'urgent need' – to hold together the nation in the midst of 'total war' – and it is in relation to this 'urgent need' that morale becomes 'something, something however that continues not to exist' (Foucault 2008: 19).

A 1941 special issue of the *Journal of American Sociology* on 'morale' exemplifies how the threat of future losses or damages to morale were brought within the state's horizon of expectation to emerge as an urgent need. Psychologist Harry Sullivan links the status of morale directly to the expansion of techniques and technologies of destruction. There is no limit or outside to war. War is everywhere. War is 'total' then because it involves a 'total' mobilisation. This mobilisation extends to the affective realm and makes morale a key resource of the nation state to be 'secured':

> The circumstances of modern warfare require the collaboration of practically everyone. Ineffectual persons anywhere in the social organization are a menace to the whole. The avoidance of demoralization and the promotion and maintenance of morale are as important in the civilian home front and the industrial and commercial supporting organizations as they are in the zones of combat (Sullivan 1941: 288).

The turn to secure the morale of a given population is intimate, therefore, with the recognition by the state of new forms of vulnerability and new ways of wounding. The special issue classifies the multiple techniques that threaten the domestic population (including elements within the population) and demonstrates how the state imagines ways in which morale could be damaged or destroyed. It is worth citing the terms used by the authors at the time to gain a sense of the catastrophic imagination through which morale is perceived to be under threat; violence functions through 'the quasi-factual', the 'ideological' and the 'analytical' (Estorick 1941: 468), the 'disorganisation of effective central control' (Sullivan 1941: 289) directed against communities, and the direct demoralisation of individuals by techniques that 'communicate a feeling of recurrent suspense, each new wave of which the victim finds himself less able to tolerate' (Sullivan 1941: 290).

Recall the role that Foucault (1980: 195) gives to an 'urgent need' in the formation and ongoing life of an apparatus. In the interview published under the title *The Confessions of the Flesh*, he gives the example of how the strategic function of the control of madness, mental illness and neurosis responded to the problem of the 'assimilation of a floating population' in a mercantilist economy (195). In the case of morale in 'total war', the anticipation of a threat to morale calls forth forms of action to prevent or prepare for it. Morale is acted on, then, in anticipation of its dissolution. The 'urgent need' is not only to ensure that a population holds together, but that the population can be mobilised to sustain and continue war and that some form of 'national character' can emerge and survive that process. In short, the 'urgent need' revolves around the population and it is the population that is threatened in the context of 'total war' and must be intervened in. How, in this context, does morale become an object-target, remembering our comments earlier about the 'existence' of the objects of apparatuses and the catalysing role of 'urgent needs' in the composition, continuation and development of apparatuses?

Let us first unpack how the population is understood, given that it is in relation to the population that morale is 'inscribed within reality' to become something. Here we see how knowing and intervening through affect can implicate other dimensions of life so that more than just affects are constituted and marked out in reality as apparatuses form, change and end.

Discussions of population in relation to 'total war' begin from an explicit understanding of population as a collective that pertains to a given territorial unit. This is either a specific area or region within the nation state or the state as a bounded geographical entity. Here the meaning of population is very close to late sixteenth- and early seventeenth-century understandings of a 'people or inhabited place' (Legg 2005). From this starting point, the population is considered to be composed of a mass of affective beings. Although morale itself remains indeterminate, as I will discuss in more detail below, it is assumed to be scored across a range of interpersonal psychological factors – such as 'combativeness, rivalry, initiative, fellow-feeling, gregariousness, docility, infectious gaiety' (Landis 1941: 332) – and the biochemical substrate of the body – such 'dehydration of the tissues of the body' (284) or 'the obscure biochemical effects which come from undercooling' (285). However, the population that makes up morale is not simply a collection of individuals grasped in terms of a preconscious, autonomic, bodily affectivity. The population is itself an affect structure. But how? There is no unanimity. Quite to the contrary, versions of the relation between collective life and morale proliferate. In one case, morale is described as a property of occasions and gatherings:

> One of the most pervasive forms in which tension and will manifest themselves in individuals and in society is in moods. Every occasion, be it a funeral or a wedding, has its characteristic atmosphere. Every gathering, even if it is no more than a crowd on the street, is dominated by some sentiment (Park 1941: 369).

In another, morale is a property of groups or associations:

> But the characteristic problems of morale belong to group temper, and it is to group mentality that the term is most characteristically applied. *Esprit de corps* is definitely a group phenomenon (Hocking 1941: 311, emphasis in original).

Elsewhere, morale is a property of a collection of 'minds' that are given the names publics or crowds:

> It is not a state of mind existing in one man alone, but in many. It is not a state of mind to be enjoyed, for itself, but to serve as a spring of action. It is not a uniform state of mind – the same under all circumstances – but is relative to the end in view (Landis 1941: 331).

The target is the population seen from one direction, its affective life. An affective life which is dispersed from the subject to, on the one hand, an affectively imbued bodily substrate and, on the other, to various types of collective. The pertinent space to act over extends from the biophysical body to gatherings or happenings.

Whilst there is no unanimity about the form of the collective, as mentioned above what is threatened is the unity or coherence of those collectives and, thereafter, how collectives are mobilised as part of 'total war'. Morale is made into a specific type of object-target through ways of thinking not only of its dissolution but the effects of its holding together. Destroying morale threatens to create a break or interruption in the life-world of a population and thus disrupt the centrifugal movement of total mobilisation. Threats to morale have a particular force, just as efforts to secure it do, because they assume a very specific relation between morale and the action of a population. Morale is the basis to action because it exceeds present diminishing affections of the body. It is a 'spring' of action (Landis 1941: 331) or a 'gift' to action (Hocking 1941: 303) because it is 'prospective' and organised around a 'faith in the future' (Park 1941: 366). Scarry (1985) hints that the basis to the promise of morale is a suggestive association between morale and the creative founding, enabling or making of future worlds. But morale is also the motive force that enables continued mobilisation under the catastrophic conditions civilians may find themselves in during 'total war': specifically conditions of 'hardship' or 'suffering' (Landis 1941: 333) in which the body is potentially affected by 'weakening influences from within (fatigue, reluctance, anxiety, irritability, conflict, despair, confusion, frustration) and from without (obstacles, aggression, rumors of disasters)' (Estorick 1941: 462).

Under a 'total' mobilisation of life and property and 'total' methods of destruction, where boundaries between civilians and the military erode, civilian bodies are exposed to a myriad of events and conditions that damage. Morale promises, therefore, to enable bodies to keep going *despite the present*: a present in which morale is either targeted directly or threatens to break given the conditions of 'total war'. And what threatens is an unpredictable, uncertain, future 'crisis' in which morale suddenly breaks or shatters, bodies are exposed to the conditions of the present, and the movement of 'total' mobilisation fails or ends. The threatening other to morale – feared by those imagining a future crisis in morale – is given a name: panic. Unlike in the early Cold War when forms of cybernetics became central to generating 'versions' of panic (Orr 2006), here panic is understood according to early behaviourist psychology as a form of *dis*organisation (see McLaine 1979). Panics differ in intensity but they are commonly understood as the dissolution of order emergent from disruptions or disturbances. These disruptions or disturbances are described in the following terms that highlight a shattering of a supposedly normal rational life: 'An event suddenly shows that the universal does not make sense and one finds one's self badly demoralized' (Sullivan 1941: 282) or in which 'any grave threat of insecurity or of cutting off all of one's satisfactions is perceived under circumstances which prohibit rational analysis and the synthesis of that wonderful thing which we call an understanding of what has happened' (282).

Destroying or damaging morale threatens to turn something interior and necessary to total mobilisation – a group of bodies, a frequently repeated activity, rational analysis, understanding – into a devastating, destructive, force (Orr 2006). An indicative 'panic provoking situation' that would generate a crisis in morale is one in which the individual as an affective being is disorganised. This is described by one of the contributors to the special issues on morale. Note how the body to be protected is described in terms of concrete visceral and proprioceptive phenomena (sensation, the skeletal system, and so on) that underpin conscious perception and deliberation:

> There will be a ghastly sensation from within, from all over within; there will be nothing remotely like reasoning or the elaboration of sentience; there will be a tendency to random activity, but practically no movement of the skeletal system because it is inhibited by diffusion of stimulus and contradictory motor impulses. As you recover, and the intense cramps which have developed in the viscera relax, you find yourself exhausted, tremulous, perhaps without control of your voice (Sullivan 1941: 279).

Unlike the disordering of panic, then, the promise of securing morale is that it enables bodies to coalesce despite the persistent presence of affections that may diminish or destroy bodies. Acting over morale offers the dream of a 'certain island of predictability' in the 'ocean of uncertainty' that is 'total war' (Arendt 1958: 220). Put differently, morale promises that the 'total' mobilisation of citizens can continue despite the excess of devastation and damage in 'total war'. It promises a means to intervene and break the relation between the capacity of an individual or collective body to be affected (through some form of diminishing encounter induced by the techniques of 'total war') and that body's capacity to affect (in this case to continue with the activities that define being a civilian in 'total war'). Morale becomes linked to 'world making' in part, then, because it is assumed to be separable from the affections of the body and, somehow, to exceed them. Morale '[t]ends to have an aura of the spiritual, to signal some capacity for self-transcendence or form of consciousness different from physical events' (Scarry 1985: 106).

Establishing morale as an object-target of power promised a way of mobilising a mass for mass destruction. It enables the otherwise unimaginable heterogeneity or bewildering abundance of modern societies to coalesce into an undifferentiated whole – a whole that thereafter acts in concert even as it resists clear and stable form. In short, morale as a property of a population is addressed as a fundamental component of a state's potential power in the inter-state dynamics of 'total war'. For this reason, securing morale becomes an 'urgent need', the problem and promise that animates the elaboration of various strategic apparatuses and around which techniques for knowing and intervening in affective life centre. For morale encourages factory productivity. Morale underpins agricultural labour. Morale sustains belief in democratic ideals. Morale powers the war economy. Morale, in short, becomes equivalent to, and substitutable for, all the other imperatives that animate the intensified concern with war during 1940s America.

As the hinge of action, morale exists as a target in 'total war' in the complicated sense that Weber (2005) argues the word 'target' originally had. Although its roots are uncertain, target probably comes from *targa*, meaning 'shield' (specifically a light and portable shield carried by archers). Remembering these defensive origins means that 'hitting' or 'seizing' a target or targets is '[l]inked to a sense of danger, to feelings of anxiety and fear, and to the desire to protect and serve' (Weber 2005: vii). The providential apparatus established on the eve of war is future-oriented in that it is animated by collective fears and anxieties that a 'crisis' in national morale is looming. The future anticipated is characterised by the inevitability of loss and damage. Suffering will happen. The question is how to deal with it. In the shadow of the catastrophic future the need to secure morale therefore becomes a necessity for the state and civil society. Morale must be mobilised if this disastrous future is somehow to be lived through and the state is to endure. Maintaining morale promises to enable 'total mobilisation' and so morale must, in turn, be secured whenever and wherever.

As we learn from Derrida (1989: 89), promises are *restless*: '[a] promise must promise to be kept, that is, not to remain 'spiritual' or 'abstract', but to produce events, new forms of action, practice, organisation and so forth' (cited in Bennett 2005). Promises call forth, demand, present action. To realise the promise of targeting morale the pre-war period witnesses the extension of an apparatus of prediction, preparedness and repair that takes the morale of collective populations as a target to be secured. Hence, it is precisely the motive power of morale that is predicted through techniques that aim to know what morale is in order to secure it, harnessed through techniques that aim to generate and maintain it, and repaired through techniques that mitigate the effects of its loss (most notably civil defence). The existence of morale, how it is inscribed within reality to paraphrase Foucault (2008), becomes inseparable from the promises that attach to securing it and the institutions that embody that promise and are set up to realise it.

At the heart of this apparatus, and the catastrophic apparatus of air power, were attempts to track the current state of morale in order to render it subsequently securable. But because of a combination of its indeterminate relation with action, and its indeterminate location, how morale shows up as an object of knowledge is similarly indeterminate. Morale becomes a vague, but actionable, 'something more' that is made ever more diffuse and ubiquitous to life. Consequently, the period before the USA's entrance into the Second World War witnessed a multiplication of techniques of measurement and calculation that attempted to track this 'something else' and thus make it subject to intervention and action.[15] In the USA this included now ubiquitous techniques for referring to and evoking affective publics such as

15 Techniques in the British context were similar. Between May and October 1941 daily reports on morale were produced which were then summarised in weekly and monthly reports by the British Ministry of Information. These were supported by mass-observation research and also drew on a variety of other sources such as postal censorship, the police

the 'social survey' (Estorick 1941) and the 'public opinion survey' (Durant 1941) that later became central to attempts to know the effects and effectiveness of aerial bombing on destroying morale (see US Strategic Bombing Survey 1947a; 1947b).

What was known through these techniques was not, however, considered to be the true nature of what morale is. Remember, morale exceeds attempts to establish it as a thing in itself. The solidity of the term object-target is potentially misleading. An object-target might be vague, indeterminate or amorphous. For example, Major James Ulio discusses how techniques for maintaining morale in the military can be used in relation to civilians, but stresses morale's excessive qualities: 'It [morale] is like life itself, in that the moment you undertake to define it you begin to limit its meaning within the restrictive boundaries of mere language' (Ulio 1941: 321). Instead, the focus for the state was on the 'conditions operative in morale formation' (Durant 1941: 413) through a measurement and calculation of the actions of the population as an aggregate of sociobiological processes. Because morale is 'like life' and 'exceeds' as a 'something more', it must be tracked indirectly. With the exception of observational methods – such as the use of mass observation (Hamsson 1976) – morale was tracked through its various and varying traces. Traces which could be found throughout life. Any aspect of life could potentially reveal the presence or absence of morale, so techniques of knowledge must know all of life without limit or remainder. Morale is everywhere. In the UK, for example, morale was understood through the frequency, extent and duration of strikes, industrial output, convictions for drunkenness or drink-driving, crimes against property (Durant 1941: 411–12). In France 'bad morale' was known through '[p]olitical tension, public violence, repudiation of existing regime by large bodies of citizens, exaggerated individualism, general passivity, demographic factors, and susceptibility to panic and despair' (408). Whilst in China morale was known through: '[d]ependence on American and British aid, the relations of the "return to the coats" school with the "new hinterland" school, the price of grain, the absence of medical facilities, and the treatment of Manchurian troops by the central government' (408).

If these lists seem arbitrary at best, they nevertheless tell us two important things about the type of object-target of power that morale becomes. First, neither morale nor these other factors are what Foucault (2007) terms the primary datum. Instead it is the interaction between the two. Morale is related to the price of grain. Morale is related to demographic factors. Morale is related to convictions for drunkenness. In short, morale varies. What techniques of knowledge do is track this movement by surveying its changes and establishing its changeability rather than simply establishing the presence or absence of morale per se. Second, and because it is 'like life itself', morale is not transparent to techniques of knowledge. It is not a stable object that can be identified and classified. Techniques of measure and calculation must engage a range of factors seemingly unrelated to morale in

and reports of W.H. Smith newsagents. From 1941 the Ministry of Homeland Security commissioned a number of surveys of public mood (see Jones et al. 2006).

order to infer its scope and effects. The result is that the governance of morale becomes the governance of life. Governance becomes 'total'. It must be found throughout life. Morale is expanded to the extent that the urban sociologist Robert Park (1941: 367) could argue that: '[w]e must recognise morale as a factor in all our collective enterprises. It is a factor in the operation of the stock exchange, quite as much as it is in the activities of the Communist party'.

The very presence or absence of morale becomes undecideable or indeterminate. It is only knowable in its many and variable effects. Here we see a connection between how an object-target is known, techniques that intervene in affective life and the promises that get attached to acting on and through affect. For this is the second sense where morale serves as a promissory note, not simply as an 'isolated island of certainty' (Arendt 1958: 220) offering the state in 'total war' its sovereign capacity to 'dispose of the future as if it were the present' (245). But further, morale becomes the horizon of intervention rather than an object available for intervention, an endlessly deferred absent-presence that can only be inferred from a seemingly arbitrary list of activities that are diffused throughout the whole of life. Paradoxically, securing morale involves the virtualisation of morale. As the target and hinge for action in 'total war', morale exists as a virtuality that comes to be known only through its varied traces. This is virtualisation in Lévy's sense of 'an "elevation to potentiality" of the entity under consideration' (Lévy 1995, cited in Weber 2004: 284). Morale is no longer an actual entity locatable in observable affective beings or affect structures through procedures of observation or experimentation. To secure morale is to elevate it to an indeterminate 'something more' that is cause and effect of an unruly excess of activities, processes and events. This elevation can be understood as a process of movement from the actual to the virtual. Thus, 'instead of being defined principally by its actuality (as a "solution"), the entity henceforth finds its essential consistency in a problematic field' (284). Morale promises, therefore, because it escapes, comes to be equivalent to life, and is thereafter absent as a delimited object. Yet, neither is morale a rare inassimilable other, akin to a punctual experience that shatters and disrupts. Instead it is commonplace, a dimension of all activities. Potentially all of life must be acted on in order to protect an exposed population, a move that echoes the emergence of a 'target rich' environment – *life* – as the object of the catastrophic state apparatuses of aerial bombing or rumour formation in which morale was taken to 'break', be 'lost', 'vanish' or 'collapse' (see Douhet 1972; Kennett 1982).

If the taking place of morale has regularities but takes on the structure of a promise, and yet 'total war' exposes all of life to new threats, then how, thereafter, can government action foster a prospective 'readiness' for action (Estorick 1941: 462)? How does acting on morale become part of the 'total control' of populations? Given that morale is under threat the key problem becomes the 'development, protection, and maintenance' of morale (Sullivan 1941: 282) and the pragmatic question becomes 'what are the methods of control by which good morale is created and preserved?' (Landis 1941: 331). But this poses a problem. Morale does not offer a graspable hinge for action. It is both everywhere, being found throughout life, and nowhere, being like life in that it escapes definition. The response is found in the development of techniques that act over the population seen from the angle of its corporeal and collective affective life, variously named as the 'collective temper of a people' (Lindeman 1941: 397), 'group temper' (Hocking 1941: 311), or 'underlying solidarity of the people as a whole' (Sullivan 1941: 300). Acting over the population involves two types of techniques. First, addressing individuals directly as affective beings primarily through forms of communication. Radio, the press, movies, theatre and educational institutions are described as the '[p]rincipal morale building agencies which are available in a democracy' (Angell 1941: 352). Second, acting over a range of factors and elements that seem far removed from morale but nevertheless implicate the collective as an affect structure. For example, how often should news of casualties be given? How can democracy move from an ideal to a fact passionately felt by subjects? How should housing be designed to ensure physical comfort? What level of physical activity should be incorporated in a national recreation plan?

Morale is not targeted, then, at the level of 'life itself', where life is understood in terms of constant self-organisation, morphogenesis and recombination. In the case of acting over morale in 'total war' we find that the activities of power resemble what Foucault (2007: 326), discussing the end of sixteenth- to eighteenth-century apparatus that in French was named police (and German, *Polizei*), termed a '[a]n immense domain ... that goes from living to more than just living'.[16] The field of action is a 'full' version of life, or as Ojakangas (2005) puts it a plenitude of life in its becoming, in which mobilisation of morale is 'total'. Because morale is scored across all life, action to mobilise morale must also occur across all of life. Consider, for example, discussion of the use of radio to generate morale as an exemplar of the intersection of techniques that act over affective beings and a population as affect structure.[17] Radio is one of a number of techniques of communication that promise to harness what Park (1941) terms the 'magical power' of morale through techniques that would enable the idea/ideal of democracy to live 'in men's minds and hearts'.

16 'Police' is discussed by Foucault as one set of political technologies from the end of the sixteenth to the end of the eighteenth century concerned with 'taking care of living' (Foucault 1994a: 413) by '[m]anipulating, maintaining, distributing, and re-establishing relations of force within a space of competition that entails competitive growth' (Foucault 2007: 312). Police is '[t]he ensemble of mechanisms serving to ensure order, the properly challenged growth of wealth, and the conditions of preservation of health "in general"' (Foucault 1994c: 94, referring to *medizinische polizei*). The fundamental object of police was the good use of the state's forces (the state's 'splendour') by acting on all the forms of 'men's coexistence with one another' (Foucault 2007: 326). Police includes everything, but seen from a particular point of view – live, active, productive man (Foucault 1994a: 412; see also 1994b).

17 Governing morale also becomes a problem of fixing and demarcating the normal from the abnormal. The one exception to the focus on 'democratic' techniques is the psychologist Harry Sullivan. Shamefully, Sullivan (1941: 294) proposed '[a] civilized version of the concentration camp' to house those who were deemed to threaten the nation's morale 'by reason of personality distortion, mental defect, or mental disorder'.

Radio is valorised for the immediacy with which it enables certain collective affects – including the warmth of voice – to be communicated at a distance. An executive at the National Broadcasting Company valorises radio along three criteria:

> (1) The immediacy of its conveyance of news; (2) the vast mass of persons thus reached, many of them having only delayed access, if any, to the newspapers, and not a few being unused to reading, or incapable of it; (3) the psychological appeal of the living human voice as contrasted with cold type – even when accompanied with the barrage of photographs now so universally employed by the press (Angell 1941: 355).

Whilst the technique addresses the individual as an affective being, a being who will be moved by the living human voice, morale is an indirect effect of other variables. Integral to the efficacy of radio is that it promises to enable a diffuse, heterogeneous, population to coalesce into a defined public that sparks into being around issues. Radio is valorised, therefore, because it promises to synchronise a heterogeneous population through the attunement of bodies at a distance. News, by contrast, is dismissed as an appropriate technique for 'developing' or 'maintaining' morale due to its relation with the innovation of newness and thus its supposed '[t]endency is to disperse and distract attention and thus decrease rather than increase tension' (Park 1941: 374).

Morale is not governed through techniques of power such as radio by establishing a direct relation of obedience or consent between a sovereign and a subject. Indeed various techniques are dismissed because they are asserted to rely on a crude 'manipulation' of morale through techniques of prescription or prohibition. It is worth comparing the use of radio as a providential technique with rumour generation as a catastrophic technique to give a sense of how 'total' methods of affective modulation work. Rumour formation was a central technique of the Office of Strategic Services Planning Group (OSSPG) 'morale operations' branch. Rumours were designed by the OSSPG to act affectively: to spread confusion and distrust, stimulate feelings of resentment and generate panic (Herman 1995). A 1943 briefing note of the OSSPG established the 'Doctrine Regarding Rumours'. It contained discussions about what a rumour was and how it worked by propagating through a population. The key question was how to enable a rumour to spread whilst retaining its original content. Properties that supposedly enabled circulation and fidelity included; plausibility, simplicity, suitability to task, vividness and suggestiveness. If this was the case then rumours could subsequently act affectively in three ways. This was summarised by the OSSPG:

> 1: Exploit and increase fear and anxiety amongst those who have begun to lose confidence in military sources.
> 2: To exploit temporary over-confidence which will lead to disillusionment.
> 3: Lead civilian populations to precipitate financial and other crises through their own panicky reactions to events (1943: 4).

Both rumour and radio act affectively. But neither guarantees to produce a direct effect or affect since both act by becoming part of the complex, living conditions that form and deform morale. The targets for rumours return us to the population understood affectively discussed above, including 'Groups or classes of people that lead monotonous lives which favour the use of fantasy' (8). Or:

> Groups or classes of people that have become fearful and anxious about their personal wellbeing. Focus on "information" that confirms the pessimistic expectations of the group involved. Extreme rumours designed to produce open panic should be timed with military action (8).

The relation of each technique – radio and rumour – to the affective surface of emergence is not simply negative. Both techniques aim to be generative of new affects of morale or panic. So rumour is designed to act by producing an 'open panic', whilst, in contrast, radio is valorised for how it may enable 'good morale' to emerge, circulate, coalesce and feed into the action of the state. In the case of radio and rumour morale cannot be brought into being directly, they are produced indirectly through techniques that become part of life. Radio and rumour both function by acting on and becoming part of the same reality as processes of morale formation. They therefore check and limit certain circulations of morale and panic by catalysing and directing others. As such they exemplify the expansion of the scope of techniques of power once the target becomes ever more diffuse. Radio and rumour are 'total' in the double sense that they extend to all of life and, to be successful and to enable the urgent need of securing morale, must become indistinguishable from life as it happens.

2.5 The Excess of Power

In the apparatuses of 'total war' morale is targeted as an indefinite potentiality in order to mobilise the affective potential of a mass formation for ongoing processes of mass destruction and mass survival (Virilio and Lotringer 1997: 24). Power is not necessarily secondary and parasitic on an insubordinate life, the potential of which power struggles to command and control. Instead, we could say that power virtualises, endlessly proliferating what should be acted on, and modulates, acting to sustain the motive power of a mass through catastrophic and providential action. Not only should we therefore take care not to assume that power reduces affective life, we should also be cautious about claims to some form of epochal change in which power only discovers affective life in the contemporary. The life of apparatuses cannot be reduced to the linear succession of periods of time, nor is there necessarily a homology between apparatuses and socio-economic transformations.

In the first section of this chapter I briefly surveyed recent attempts to diagnose how power now works through affective means, principally, although not exclusively, through forms of manipulation that work on autonomic bodily reactions (see Barnett 2008). In distinction from arguments that maintain affect is an object-target of new modalities of power, I would stress that affective life has long been an object-target intertwined with various modes of power that coexist, resonate and interfere with one another (rather than necessarily replace one another in a relation of linear succession). In this chapter these modes of power have included forms of sovereign power (of the state in 'total war') and forms of vital power (in how radio functions, for example). There are, have been and will be others; not least the forms of sovereign and environmental power to be described in Chapter 3 in relation to spaces of torture as the War on Terror morphs into a global counter-insurgency. The question of the status of affect 'today' is, then, a matter of understanding specific modes of power, their changes and the complicated relations between them. Whilst the current interest in affective manipulation has been placed in the context of the emergence of a politics of 'life itself' (Hardt and Negri 2004), the case of morale opens up a longer genealogy of interventions that implicates a wider range of knowledges, including in this case social psychology and cybernetics. We should take care, then, not to reproduce a type of presentism in analyses of the politics of affect that would assume the novelty of affective manipulation.[18] We should also avoid any model that presumes that, outside of the contingency of demands, mechanisms and functions, there is something like a transhistorical constant to object-targets such as morale (Foucault 2008: 63).

Morale in a state of 'total war' is but one example of how forms of power/knowledge constitute affects and other referent objects, specifically in this case the population, as particular kinds of collective affect structures. From the example, we get a sense of how targeting and securing morale emerged in response to an 'urgent need' – the need to sustain the population's participation in the activities of war – and how the apparatuses of 'total war' involved the integration of multiple elements: surveys, speculation, radio waves, and so on. What the case also allows us to consider is the precise relation between apparatuses and affect. When discussing how affective life is known, targeted and acted upon, there is a tendency to presume that this is a reductive process of abstraction and limitation. The case gives us cause to pause and rethink what happens what affective life is rendered actionable through forms of knowledge and is subject to intervention. For the mechanisms discussed in the example of targeting morale do not simply reduce the excess of affect through some form of capture or control. Rather, they result, in different ways, in efforts to know and act on affects as collective

18 The immediate context for Foucault (2008: 63) is his discussion of freedom in the lecture of 24 January 1979 in which he stresses that freedom must be understood in a non-substantialist manner: 'we should not think of freedom as a universal which is gradually realized over time, or which undergoes quantitative variations, greater or lesser drastic reductions, or more or less important periods of eclipse. It is not a universal which is particularized in time and geography. Freedom is not a white surface with more or less numerous blank spaces here and there and from time to time'.

phenomena intimate with life's indeterminacy. An indeterminacy that is itself indeterminately located, pertaining to both the affective substrate of bodily life and to how populations form and deform. Even though it provides a less catchy starting point, we could say that not only do we not know what a body can do, we also do not know what apparatuses and modes of intervention can do. Instead, we must pay attention to the specificities of the apparatuses through which affective life becomes an object of thought, including the causal role of 'urgent needs', the ways in which a named affect is 'inscribed within reality' (Foucault 2008: 20) as an object-target, and how forms of power are enacted, expressed and reflected as affective life is intervened in.

Focusing on the apparatuses through which affective life is known, rendered actionable and intervened in implies that object-targets change, but that change should not be thought of through a model of linear replacement and succession. In the next chapter I address the question of how object-targets change by turning to consider the life of apparatuses: that is, how apparatuses emerge, coalesce and change and how multiple 'versions' (Despret 2004) of affect are enacted. I do so by way of a story of an affective state first named in the early 1950s: 'debility, dependency and dread'.

Chapter 3
Versions

> The lines in the apparatus do not encircle or surround systems that are each homogeneous in themselves, the object, the subject, language, etc. but follow directions, trace processes that are always out of balance, that sometimes move closer together and sometimes farther away. Each line is broken, subject to *changes in direction*, bifurcating and forked, and subjected to *derivations*. Visible objects, articulable utterances, forces in use, subjects in position are like vectors or tensors
>
> (Deleuze 2008: 338, italics in original).

3.1 Debility, Dependency, Dread

Two bodies and two environments. Environments that have been set up to produce a specific affective state in those that dwell, forcibly or by choice, within them. The affective state has a name: 'debility, dependency, dread'.

The first environment is inhabited by a now unknown male American high school student. He has volunteered to take part in a trial at McGill University in Montreal, Canada, in the early 1950s. The trial was run by a Dr Donald O. Hebb. It is described in the proceedings of a symposium held at Harvard Medical School on 20 June and 21 June 1958:

> Male college students were paid to lie 24 hours a day on a comfortable bed in a lighted, semi soundproof cubicle which had an observation window. Throughout the experiment, the students wore translucent goggles which admitted diffuse light but prevented pattern vision. Except when eating or at the toilet, they wore cotton gloves and cardboard cuffs which extended from below the elbows to beyond the fingertips, in order to limit tactical perception. A U-shaped foam rubber pillow, the walls of the cubicle, and the masking noise of the thermostatically regulated air-conditioner and other equipment severely limited auditory perception (Heron 1965: 8).

The second body is of Shafiq Rasul, a British detainee in the 'War on Terror'. He is one of three friends dubbed the 'Tipton 3' by the British print media who were held for 29 months at Guantánamo Bay, before being released without charge. After being released in March 2004, Shafiq bore witness to the shattering experience he endured, including the following that occurred 'around the end of 2002' (Shafiq Rasul quoted in Otterman 2007: 152):

> I was taken into a room and short shackled. This was the first time this had happened to me. It was extremely uncomfortable. Short shackling means that

> the hands and feet are shackled together forcing you to stay in an uncomfortable position for long hours. Then they turned the air conditioning on to extremely high so I started getting very cold. I was left in this position for about 6 or 7 hours, nobody came to see me. I wanted to use the toilet and called for the guards but nobody came for me. Being held in the short shackled position was extremely painful but if you tried to move the shackles would cut into your ankles or wrists. By the time that I was eventually released to be taken back to my cell I could hardly walk as my legs had gone completely numb. I also had severe back pain (152).

What links the two bodies and two rooms across around 50 years is an affective state and a set of techniques designed to induce it. First named in 1957, it is the term given by behavioural psychologists to the experience of the anonymous American student and by the US military to the experience of Shafiq Rasul. What distinguishes the two bodies is their relation to that state. The unknown male is the subject of an experiment. He stops the experiment before the planned time, unable to bear it. For Shafiq Rasul, there is no such choice. He remains subject to techniques designed to induce 'debility, dependency, dread' and other affective states intentionally produced at Guantánamo Bay until his release.

Bodily capacities are constituted through attempts to intervene in and through affective life. In the case of morale, rumours were supposed to engender panic. Radio was supposed to gather together people around a common purpose. Life is saturated with attempts to intervene in what bodies can do. From the designing of retail atmospherics, through to a whole pedagogy of self-help, affects are intervened in, moved and changed. Affective life is 'informed' (Barry 2006) by attempts to know and act on affective or emotive experience. Perhaps, though, 'debility, dependency, dread' is the paroxysmal point of such interventions because what is produced through 'debility, dependency, dread' is an absence – the absence of a subject who can relate to his or her environment and make sense. Instead, people are stripped of everything that makes them subjects. Unlike other experiences, such as the thrills of sports, the ecstasies of drug use or the vicarious pleasures of video gaming, in 'debility, dependency, dread' individuals '[h]ave reduced viability, are helplessly dependent on their captors for the satisfaction of many basic needs, and experience the emotional and motivational reactions of intense fear and anxiety' (Farber, Harlow and West 1957: 273).

This chapter follows how 'debility, dependency, dread' was known, named and became a means of damaging people in apparatuses of violence. As well as bearing witness to the weaponisation of an affective state, I also want to focus on how the apparatuses in which affects are object-targets emerge and change and the effects this has on how life is known, rendered actionable and intervened in. For 'debility, dependency, dread' is not the same object-target in the experimental spaces of 1950s behaviourism as it is in the global prisons of the War on Terror. It is articulated with other affects in different ways. It is linked to the subject in different ways. It is produced for different purposes. And it is experienced in different ways. In other words, it cannot be assumed that an object-target is a single thing that remains consistent across the 'lines' (Deleuze 2008: 338) that make apparatuses. My emphasis in this chapter is, then, on changes to an object-target. This is to complicate further how an object-target of apparatuses exists, or takes on existence. If in the previous chapter I argued that object-targets may be indeterminate, then in this chapter I focus on how object-targets are in process, constantly being formed and transforming. What is produced across apparatuses are multiple partially connected 'versions' (Despret 2004) of what an affect is.[1] Despret (21–36) introduces the term 'version' in order to understand the coexistence of alternative accounts of the emotional or affectual without explaining away that difference as a result of different interpretations of emotion/affect, or as a result of a distinction between the true nature of emotion/affect and various secondary illusions or errors. As with the 'after actor-network theory' literature on multiple enactments of a phenomenon (see Mol 2002), what is useful about the term version for my purposes is that it emphasises multiplicity and therefore allows us to attune to how versions relate to one other. For, as Despret (30) puts it, 'versions coexist in the same world and are cultivated in that world'. Those versions do not relate back to an ahistorical or ageographical universal. Instead, versions are entangled with other versions existing in other spaces or times. Every version makes a world exist in a possible manner, brings something into existence, and holds together alongside other entities and relations (31). A version of morale or 'debility, dependency, dread' is constituted alongside other entities.

In the first section I turn to the naming of an object-target – 'debility, dependency, dread' – that is in many respects the counterpoint to morale. If morale promised to enable action to continue into an indefinite future, 'debility, dependency, dread' is the absence of action in an eternal present. If 'debility, dependency, dread' involves the forced removal of subjectivity, morale holds together collectives. Tracing the naming of this state in 1950s behaviouralism, and in response to a structure of feeling that we can term 'Cold War paranoia', I draw out how object-targets change as the apparatus within which they are known and acted on changes. Section two steps back and supplements the concept of apparatus discussed in the previous chapter in order to think through how objects-targets are enacted, or in Foucault's (2008: 20) terms how they are 'inscribed in reality'. Returning to the example in the following section and moving to the shameful histories and legacies of torture in the Cold War and War on Terror, I explore how versions change through various techniques of damage and destruction linked to the elaboration of mechanisms and functions in relation to new 'urgent needs'. Just as the previous section gave us a sense of the life of apparatuses, here we get

1 One example Despret (32) traces of a version of emotion is a now familiar one: 'the one that gives us passions as internal, reactive, biological, universal, and natural'. Of course, it is this version of emotion that most recent work on affect and emotion in geography and other social sciences has been defined against, even as elements from it have been recuperated.

a sense of the liveliness of versions such as 'debility, dependency, dread' and how change may happen without assuming the linear succession of epochs.

In conclusion I reflect on the first form of mediation the book is organised around: apparatuses that know, render actionable and intervene in affective life. Here I explore how critique might function as a specific practice of hope that aims to disclose turning points in the composition and coordination of apparatuses.

3.2 Keep Hope Alive

In a paper published in the psychology journal *Sociometry* in 1957, an affective state is given a name – 'debility, dependency and dread'. Farber et al. (1957) place their article in the context of the anxiety that in the early 1950s surrounded what was variously called 'brain washing', 'thought control', 'thought reform' or 'menticide'.[2] The problem was that captured American servicemen confessed to fantastical crimes or publically professed support for communism. Various hypotheses were offered as to how and why this had happened. Was it because of the moral laxity of post-war society? Were the Communists using a special type of truth serum? Had the Communists discovered a hitherto undiscovered way of manipulating the 'minds of men'? In the context of a hyper attentiveness by the US as to whether there really was a 'balance of powers' between the US and USSR, Farber and colleagues made a novel move. They deny that there is anything mysterious or particularly sophisticated about Communist 'brainwashing'. In other words they deny an explanation that would impute a mysterious power to Communist governments, a power unavailable to Western capitalist states, and a power that threatened those states by undermining Western individualism. For Farber et al., 'brainwashing' was an explainable phenomenon: an effect of the Communists' environmental manipulation. Specifically, the:

> '[e]xtraordinary degree of control the Communists maintain over the physical and social environments of their prisoners (Farber et al. 1957: 271).

In the midst of a mood in the US that historians of the era have variously described as 'cold war paranoia' or 'fear of communism', identifying 'debility, dependency, dread' becomes a way of understanding the effects and affects of a peculiar type

2 The study was based on a report for the 'Study Group on Survival Training', which was sponsored by the Air Force Personnel and Training Research Center. The 'Study Group' was set up in the context of efforts to explore protection against 'brainwashing' in the context of fears over Communist 'brainwashing' (such as the press conference given on 26 January 1954 at Panmunjom by 21 American POWs). The resulting 'stress inoculation' programme aimed to 'condition' soldiers to resist brainwashing methods. The emphasis on 'conditioning' resonates with the use of forms of air power to 'condition' subjects on the ground (see Adey 2010).

of 'total' environmental control; one based on a degree of control supposedly unavailable in an 'open society' that, at the time, equated freedom with particular forms of individuation based on intentionality and the capacity to deliberate, decide and initiate action.

What Farber and colleagues do is name this affective state and its component experiences. They do so by reference to experimental work on sensory deprivation, first person accounts by returning prisoners of war, debrief sessions with those same prisoners and what can best be described as guesswork. Here is how they characterise 'debility, dependency, dread'. Note the caveat they offer in terms of the different intensities of 'debility, dependency, dread' and the striking importance they grant to 'environmental conditions':

> Although the specific components of these states vary in intensity and pattern, in the case of the prisoner of war they contain at least three important elements: debility, dependency, and dread. They refer to the fact that individuals subjected to the kinds of environmental conditions listed above have reduced viability, are helplessly dependent on their captors for the satisfaction of many basic needs, and experience the emotional and motivational reactions of intense fear and anxiety (272–3).

Dread, debility, and dependency are described as separable 'elements' of the affective state that nevertheless 'interact' to produce an overall experience that shatters the subject. These are all affective, albeit in different ways. Consider 'dread', described in the following way:

> Dread is the most expressive term to indicate the chronic fear the Communists attempted to induce. Fear of death, fear of pain, fear of nonrepatriation, fear of deformity or permanent disability through neglect or inadequate medical treatment, fear of Communist violence against loved ones at home, and even fear of one's own inability to satisfy the demands of insatiable interrogators – these and many other nagging despairs constituted the final component of the DDD syndrome (273).

The presumption that the subject is an affective being is also in the background to 'debility' and 'dependency'. The former is described in terms of physical bodily states (disease, semi-starvation, fatigue) that led to a 'inanition and a sense of terrible weariness and weakness' (273). The latter is described in terms of a reliance on the captor produced by prolonged deprivation of 'many of the factors, such as sleep and food, needed to maintain sanity and life itself' (273) together with 'occasional unpredictable brief respites, reminding the prisoner that it was possible for the captor to relieve the misery if he wished' (273).

What is diagnosed at the root of the phenomenon of brainwashing is the dissolution of a particular type of subject: the liberal subject with an internal life. By virtue of a total manipulation of the environment, the subject has lost most of the capacities for action that makes him/her into a subject in the first place. Named consequences include: a 'condition of markedly reduced responsiveness', 'reduced energy', 'the frustration of previously successful techniques for achieving goals', 'disruption of the orderliness, i.e. sequence and arrangement of experienced events, the process underlying time spanning and long-term perspective'. Ultimately:

> By disorganising the perception of those experiential continuities constituting the self-concept and impoverishing the basis for judging self-consistency, DDD affects one's habitual ways of looking at and dealing with oneself (275).

The effect is likened to 'postlobotomy syndrome' (275), 'schizophrenia' (275) or the 'hypnotised subject' (278). That is, the state takes on its meaning by virtue of the comparison Farber and colleagues make with what they consider to be abnormal affective states. What links them all is that the subject is no longer a liberal subject as then understood. The subject's internal life – their capacity to reason and feel – becomes a secondary effect of the 'total manipulation' of an environment. At the heart of such behavioural explanations was a reversal to the standard way in which the relation between the subject and environment was framed. Rather than possessing an interior life, the subject is made an outcome of the environment that he or she dwells within and is surrounded by. Capacities to act, even the motive force of interest, become contingent on there being an environment that fosters and sustains (through reward or punishment) appropriate behaviours. What matters is the environment and the subject's responsiveness to it. Subjects are made into secondary effects of an environment of which they are but one part existing alongside others. Compare with the case of morale in a state of 'total war'. There what was acted on was the national population understood as a unified affect structure, rather than the environment understood as that which surrounds and enables the subject.

The promise of Farber and colleagues' version of 'debility, dependency, dread' was that if brainwashing was an effect of 'debility, dependency, dread' and 'debility, dependency, dread' was an effect of environmental manipulation, then 'brainwashing' as a weapon of war could be responded to and controlled. After the failures of experiments with psychotropic drugs and other ways of understanding the phenomenon of why soldiers would confess to what they had not done, understanding 'brainwashing' through behaviouralism seemed to offer a hope: a way of counteracting the actions of the enemy through 'environmental control'. As discipline acts on capacities and biopolitics populations (Foucault 1978), behaviouralism shifts attention to the environments within which the subject dwells, in doing so constituting a new object-target: individual-environment. What is presumed by behaviouralism is that the subject is not an isolated individual but instead responds systematically to modifications in her/his environment (after Foucault (2008: 259–60, 260–61n) on 'environmental technologies').[3] An 'environmental technology' presumes that; (a) all people are responsive to 'to some extent' (259) to the possible gains and losses of an action and; (b) that instead of working through a 'general normalization' or 'normative mechanisms' action is brought to bear on the environment, or what Foucault (260) terms 'the rules of the game rather than on the players'. By making brainwashing a matter of milieu control, Farber and colleagues not only describe 'debility, dependency, dread' but they also render it actionable because of their identification of the environment as a distinctive object-target to be known, rendered actionable and intervened in.

Farber and colleagues' explanation of 'debility, dependency, dread' was both to downplay the power of 'brainwashing' and to make systematic control available to others. It was not because of some mysterious power the Communists possessed that they were able to force confessions, but through what Farber et al. term the systematic application of 'normal principles of behavior'. Their claim that 'debility, dependency, dread' is produced through a manipulation of the 'environment' follows from a reworking of behaviouralism through information theory and cybernetics, in particular ideas of feedback systems. The individual subject can be intervened in through his or her 'environmental conditions' because those conditions were rendered actionable as complex information systems. Specifically, brainwashing follows from the ability behaviouralism offered to manipulate a new referent-object: the environment or conditions, understood as an information system:

> Although the behavior of some prisoners under Communism, including collaboration, conversion, and self-denunciation, appears to suggest that Communists are able to "brainwash" their prisoners in a mysterious way, a consideration of the physical, emotional, and social conditions of the prisoner in conjunction with the ordinary principles of human behavior reveals that such behavior may be readily explained (282).

Once described on the basis of reports of captured airmen and the types of experimental studies discussed above, the research team must thereafter understand how the state can be produced and countered. But their attempt is speculative. For the key problem that is being investigated is not, importantly, the production of 'debility, dependency, dread' per se, it is brainwashing. And brainwashing is understood not as a function of 'debility, dependency, dread' but as an effect of the alleviation of 'debility, dependency, dread'. This is important. It is the occasional

3 In brief, but as ever suggestive, comments in the *The Birth of Biopolitics* series (principally the lecture of 21 March 1979), Foucault (2008) ties the emergence of 'environmental technologies' to neo-liberal forms of governmentality. The ideas are suggestive and demand that we are specific in identifying the relation between ways of representing and intervening in affective life and modalities of power.

mitigation of 'debility, dependency, dread' that is key against the backdrop of an environment that has been manipulated to produce that affective state:

> Relief, whether due to spontaneous factors or deliberate manipulations, is intermittent, temporary, and unpredictable. Far from weakening the expectancy of relief, however, this tends to maintain the expectancy and renders it less susceptible to extinction, In nontechnical terms, this process seeks to *keep hope alive*, permitting some degree of adaptive behavior, and inhibiting self-destructive tendencies (276, emphasis in original).

'Debility, dependency, dread' is not hopelessness. In fact it must not become hopelessness if it is to work. Rather it works as the 'occasion for the selective reinforcement of certain modes of response' (276) because hope is kept alive: hope that the prisoner's distress may be relieved. For Farber and colleagues, it is the alleviation of the state that leads to the subject learning various instrumental acts. Moreover, if the alleviation is unpredictable – rewards are given for various behaviours without a pattern that is discernible to the prisoner – then the subject will give the maximum number of responses and be the most compliant. What is important to understand, then, is the oscillation between 'debility, dependency, dread' and other affective states that bring relief from that state. It is by manipulating 'relief' that the affective subject is acted on:

> Relief of hunger, fatigue, isolation, or pain, even temporarily, serves as an automatic reward. Even the verbal and empty promise of alleviation of DDD leads to appropriate anticipatory goal responses, keeping hope alive. Paradoxically, interrogation, harangues, threats, and contumely may also have a rewarding aspect, so great is the acquired reinforcement value of social communication and speech under conditions of isolation, dependency, and physical debility (277).

Brainwashing is inseparable, then, from a set of techniques that manipulate the environment in order to keep the prisoner's hope alive. If hope remains the prisoner will change his behaviour to avoid or alleviate 'debility, dependency, dread'. If hope is lost then the prisoner will become uncooperative. Farber et al. have identified the object-target of brainwashing: hope. And in common with the emergent 'environmental technology' of behaviouralism they have articulated a new object-target for intervention: the environment understood as an information system. The response to the problem of brainwashing is inseparable from a specific version of affect. Affect is rendered actionable by controlling the environment.

3.3 Materialisations

One common element within the apparatuses that have the 'capacity to capture, orient, determine, intercept, model, control, or secure the gestures, behaviors, opinions, or discourses of living beings' (Agamben 2009: 14) are representations of what affect is and what specific affects are and do. For this reason, ordinary affective life is informed by representations of affect and emotion. Numerous descriptions of what affects are and do coexist, resonating or contradicting one another, and becoming part of apparatuses. 'Debility, dependency, dread' is produced in one particular apparatus involving, amongst other elements that come together, specific ways of knowing subjects as affective and environmental beings. In particular, it comes freighted with the institutional position, prestige and emerging power of a combination of late 1950s behavioural psychology and cybernetics and information theory.

What function do representations of affective life have in the apparatuses through which ordinary affective life is patterned and organised? To begin with let us reject any simple separation between representation and affect. A separation that would postulate a gap to be bridged between an ontologically distinct realm of flowing intensities and practices of representing and intervening in affect (where affect would always-already be in excess of representation). If in the wake of social constructionism word has held primacy over world (Barad 2003), then recent work on affect has given the opposite impression – that affect names a domain of unmediated experiential richness separate from practices of representation. We have taken the first steps in the previous chapter to think a different relation by identifying apparatuses through which affective life is known and rendered actionable. Now we need to develop the emphasis on the life of apparatus with a more detailed account of how distinct versions (Despret 2004) emerge and change. I am helped in this task by work in feminism and science studies[4] that has begun thinking about the materiality of knowledge practices and moved toward rethinking representation as a set of practices that make, unmake and remake (for example Barad 2003; 2007). Like Despret's (2004) work on versions of emotions and existing in complex relation to theories of performativity, what this work does is understand versions of affective life as material consequences of diverse, technically mediated, explorations in affective life.

Breaking with the sterile opposition between social constructionism and naturalism, this work begins from a performative understanding of representation. Summarising too crudely, this involves two starting points. First, representations do not serve a mediating function between knower and known. The question is no longer whether a representation of an affective state is accurate in relation to an independent reality or how reality is veiled by a representation. It is instead

4 Echoing various non-representational theories, for Barad (2007: 41) 'representationalism' is 'the belief in the ontological distinction between representations and that which they purport to represent; in particular, that which is represented is held to be independent of all practices of representing'. As Barad (49) goes on to make clear, what performative approaches share is that inquiry is focused on 'the practices or performances of representing as well as the productive effects of those practices and the conditions for their efficacy'.

a question of paying close attention to the composition of actual practices of representing and the way in which versions are therefore emergent from entangled intra and inter relations (after Barad 2007). Second, representations do not somehow inevitably reduce the uncontainable mystery of 'affect itself'. Affect is not some kind of ungraspable exteriority that representation can only fail in relation to. Instead representations are themselves active interventions in the world that may carry with them or result in changes in bodily capacity or affective conditions. The focus here is on how representational forms – concepts, words, graphs, diagrams, and so on – are elements within apparatuses and, as such, are integrated with other elements to mediate versions of specific affects. Representations are presentations that create worlds rather than representations of some pre-existing order. They have an expressive power as active interventions in the fabrication of worlds and are integrated alongside other discursive and non-discursive elements in apparatuses. Dewsbury et al. (2002: 438) put this well when they stress the need to take the force of representation seriously: 'representation not as a code to be broken or as an illusion to be dispelled rather representations are apprehended as performative in themselves; as doings. The point here is to redirect attention from the posited meaning towards the material compositions and conduct of representations'. To misquote the affirmation of a body's openness that Deleuze finds in Spinoza, before focusing on this or that apparatus we do not know what representations can do.

This account of representation is unusual, then, in being thoroughly materialist, in the sense of attending to how representations make a difference. To paraphrase Doel (2004), nothing is spirited away into a separate realm of representation. There is no distinction between dead matter and enlivening signification (see Doel 2010; Dewsbury 2010). Everything may act, including the images, words, reports and texts that inform affective life (Laurier 2010). The question is a pragmatic one of how and with what effects?[5] Let us return to the example of 'debility, dependency, dread' to think through the consequences of approaching representations of affect as 'small cog[s] in an extra-textual practice' that involves the gathering of heterogeneous elements (Deleuze 1972 in Smith 1998: xvi).

First, 'debility, dependency, dread' becomes an entity by being named through a set of practices of representing that are integral to the ongoing (re)elaboration

5 Here we could draw a crude contrast between two different theories of representation found in non-representational theories (both of which break with understandings of representation as something that mediates between a subject and the natural world). First, the version I am outlining here: that in effect gives no special status beforehand to representations. As Deleuze put it, they are one element amongst others in an extra-textual practice. The inspiration here comes from approaches that focus on what Philo (2011) has termed 'discursive life' (see Laurier 2010). Second, are slightly different approaches that take representation to be a form of calculation that captures and closures the world, whilst also always-already leaving a trace or remainder that betrays the failure of representation (see Harrison 2007; Dewsbury 2010).

of a behaviouralism-war apparatus. It is made knowable through interviews, observational studies, experimental studies, reports and laboratory instruments. In short, it is made knowable through an apparatus that is itself a 'dynamic (re) configuring of the world', as Barad (2003: 816) puts it. These practices of knowing interact with others that are also attempting to explain the phenomenon of brainwashing. Understanding it in terms of 'debility, dependency, dread', replaces an explanation that imputes brainwashing to the use by Communist countries of hitherto unknown psychotropic drugs, for example. The apparatus involves but exceeds elements from behaviourism. The notion of 'environment' is rendered equivalent to the notion of 'conditions' in behaviouralism. By virtue of being named, 'debility, dependency, dread' becomes an element within the apparatus of experimental science without being reducible to it. Naming an affective state is a referential act in that naming presumes the pre-existence of the entity named. The experience of 'debility, dependency, dread' would, on this understanding, pre-exist the act of naming. Of course the name would fail to capture fully the unspeakable experience of being 'brainwashed', but it would nevertheless house that experience in 'relatively stable nominal unity' (Caputo 2006: 2). Naming is also an evocative act in that it gestures towards something which exceeds the name. In this case the smooth alliteration evokes an experience that undoes the subject. 'Debility, dependency, dread' hinting towards an abhorrent experience; conjuring spectres of bodies lying prone, hinting darkly to a history of experimental science and techniques and technologies of 'milieu control' through cybernetics and information theory aligned to behaviouralism.

Second, the representation of the affective state itself has an affective life. It contributes to a hope that Communist brainwashing is an explainable phenomenon that can be countered and, in time, perhaps replicated. It contributes to a hope in the institutions set up to counter the then quasi mystical 'brainwashing', bolstering the institutional power of a then still nascent behaviouralism by aligning it with information theory and cybernetics. It does not just reorder bodies but reorders other material phenomena. Interrogation policy gets changed, spaces constructed, nations denounced, all in relation to the phenomenon of 'debility, dependency, dread'. What is produced is a version: an effect with boundaries, properties and meaning that is composed through a set of interacting components that cannot be separated (Barad 2007). A version is *both* a name with a material referent in a specific affective experience and is itself a new entity that has an affective life of its own. 'Debility, dependency, dread' becomes inseparable from the threat of war and the promise of defeating an enemy as it moves from the laboratories of American universities to the spaces of suffering that mark the so-called War on Terror.

On this account, 'debility, dependency, dread' takes on existence in the space between a set of material configurations of the world associated with a behaviouralism aligned with cybernetics and a set of specific material phenomena. This opens up two further questions. How do representations of affective life have their own historicity, being themselves elements within apparatuses, being themselves geo-historical forces? Following on, how do the apparatuses within which representations of affect take on their meaning and efficacy themselves change as representations are translated into techniques that (re)make capacities?

Let us return to the case of 'debility, dependency, dread' but leap forward some 50 years to a different war. From American service men confessing during the Cold War to Iraqi men and boys being forced to listen to music from a children's TV show that I used to watch with my now four-year-old daughter, Edith.

3.4 'I Love You. You Love Me'

'I love you. You love me. We're best friends like friends should be'[6] The favourite song of the popular children's TV character Barney was also one of the US military's favourite songs to play at loud volumes to detainees in Iraqi. Over the course of days of captivity, the song would be played thousands of times at ear splitting volumes to detainees otherwise forcibly deprived of sensory stimulation. When news of Barney's participation in torture broke in the British *Guardian* and other newspapers, the story was originally greeted with mild amusement and not a little bemusement, originally breaking in 2003 before the full horror of what was happening at sites of interrogation and detention became publicly known (after which it becomes evidence of the US government's torture). Jokey parallels were drawn with the childcare experience of US and UK parents. For parents, the Barney song was a mildly irritating yet harmless accompaniment to everyday parenting. In Iraq the Barney song was used along with other music as one sonic element within an environment designed to 'break' interrogation subjects.[7] The specific purpose of its repetition was to induce 'disorientation' in detainees. Barney's sugar-coated song of love and friendship was weaponised as part of military psy-ops, used as what is termed in PSYOPs and interrogation field-manuals as 'futility music' designed to enable a total control of the environment of incarceration, break subjects and render them susceptible to interrogation.[8]

Only provisionally stabilised in the apparatuses of early 1950s behaviouralism, 'debility, dependency, dread' has gone through a number of translations in other

---

6 The song is from the children's TV series *Barney and Friends*. The show features a purple dinosaur named Barney. *I Love You* is Barney's theme song.

7 Songs included *Fuck your God* by the heavy metal band Deicide, Metallica's *Enter Sandman* and David Gray's *Babylon*. Selections of songs appear to relate more to the idiosyncrasies of soldiers tasked with interrogation than any centralised policy. There is of course nothing new in the use of music in torture or warfare. We might think, for example, of the US military's use of loud music to induce Panamanian president Manuel Norriega's surrender and the efforts to develop acoustic weapons (see Cusick 2006). Whilst from sonic booms to the roar of aeroplanes, music, noise and sound have long been weapons of war. On music/sound as weapon see Goodman (2009).

8 For a comprehensive account of the use of 'futility music' in spaces of detention as part of the 'global War on Terror' see Cusick (2008). See Cusick (2006) for an overview of 'acoustic bombardment' and 'acoustic weapons' in the battlefields of Iraq.

apparatuses to reach the stage where people are damaged through the repetition of a song about a purple dinosaur. It is an experience – if that is the right word given that the aim is to damage the subject who experiences – akin to 'debility, dependency, dread' that the US military use 'futility music' to produce. Indeed there is a line of descent from the explication of environments in the original Cold War apparatus to the playing of the Barney song to create unbearable environments in Iraqi prisons. Most notably for the so-called War on Terror, 'debility, dependency, dread' was named and articulated in the CIA's 1963 manual *KUBARK Counterintelligence* (KUBARK being the CIA's codename for itself). As a number of academic and popular accounts have now shown, many of the practices of 'coercive' and 'non-coercive' torture undertaken by the USA mimic those suggested in the 1963 manual (see Danner 2004; McCoy 2006). As such, the manual becomes a key element that moves between, and is redeployed across, the partially connected military apparatuses that fold Cold War practices into the War on Terror, blurring the lines between the two wars.[9] In the manual, the subject is still defined as an environmental being and it is by acting on the environment that the subject can be changed and intervened in, manipulated and moulded. Unlike the collective subjects of morale I discussed in the previous chapter, the 'environmental' subject can be acted upon because he/she is susceptible to manipulation of the environment. The environment therefore becomes the point of contact between interrogator and interrogatee, the point of contact between the subject and a specific mode of power based, in part, on a configuration of sovereign violence in the midst of a new type of war.[10]

How does 'debility, dependency, dread' as a version with determinate properties, meanings and boundaries change across apparatuses? Most simply, it becomes an affective state to be intentionally produced in a different relation of power. No longer linked to a relation of experimentation, the scene of interrogation involves an interaction characterised by an asymmetric power relation. The environment is manipulated by the interrogator for his/her purposes. In the context

9 The 1963 KUBARK manual is the source material for the also declassified *Human Resources Exploitation Training Manual 1983* (the 1983 manual also included material from what was known as Project X – a military effort to generate training guides for Latin America based on the experiences of counter-insurgency techniques in Vietnam). Whilst the 1963 manual was related to Communist subversion generally but the Soviets specifically, the 1983 manual was deployed to chilling effect in Columbia, Peru, El Salvador, Guatemala and Ecuador in the 1980s, being used in at least seven US training courses and being propagated through The School of the Americas (see McCoy (2006) on how these methods were propagated through Latin America).

10 Given how brief his comments are on 'environmental technologies' it is unsurprising that Foucault (2008) does not link environmentality to sovereign power. He hints, of course, that any life which does not or cannot be made to fit with the market is cast out and either made to die or left to die, but here I am pointing to something a little different: a form of power that stands in-between environmentality and sovereign power by investing the body and marking it but through the environment rather than directly.

of this relation 'debility, dependency, dread' becomes the desired product of an 'environmentality' linked to a sovereign relation of power over the body. Drawing extensively from Farber et al. (1957) as well as other behaviourists, the KUBARK manual advocates a set of 'coercive' and 'non-coercive' interrogation methods. These are the means to control the environment. More specifically, the means of inducing 'the debility-dependence-dread state' is a 'a homeostatic derangement', a disruption to the ordinary environment which enables the liberal subject to subsist. What marks out the 1963 manual, in distinction from Farber et al., is that the 'debility-dependence-dread state' is understood as a form of regression – where the idea of regression introduces a version of affect tied to psychoanalysis.

The focus on 'regression' is described by way of Hinkle's (1961) study 'The Physiological State of the Interrogation Subject as it Affects Brain Function'. Regression is a loss, a stripping away of what makes the subject a subject, and a return to a life separated from the prevailing habits and norms of action and reaction. Quoting Hinkle, the manual describes regression in the following terms:

> The result of external pressures of sufficient intensity is the loss of those defences most recently acquired by civilised man. " ... the capacity to carry out the highest creative activities, to meet new, challenging, and complex situations, to deal with trying interpersonal relations, and to cope with repeated frustrations. Relatively small degrees of homeostatic derangement, fatigue, pain, sleep loss, or anxiety may impair these functions" (KUBARK 1963: 83).

'Homeostatic derangements' are described in strikingly affective terms. Fatigue, pain, and so on, are all 'induced' by the 'control' of the subject's relation with his/her environment. Pain follows from a discomforting environment, for example. Fatigue from an environment in which sleep is forcibly interrupted or foreclosed. What is created is an 'interval of suspended animation' in which the subject loses his/her normal capacity to function and is therefore rendered susceptible. This 'interval' is different from the 'hope' that Farber et al. argue is vital to 'keep alive' if interrogation is to work. The 'interval' is a moment of shock; a point of suspense after environmental interventions have undone the subject but before the subject has been utterly destroyed:

> The effectiveness of most of the non-coercive techniques depends upon their unsettling effect. The interrogation situation is in itself disturbing to most people encountering it for the first time. The aim is to enhance this effect, to disrupt radically the familiar emotional and psychological associations of the subject. When this aim is achieved, resistance is seriously impaired. There is an interval – which may be extremely brief – of suspended animation, a kind of psychological shock or paralysis. It is caused by a traumatic or sub-traumatic experience which explodes, as it were, the world that is familiar to the subject as well as his image of himself within that world (65–6).

The moment of shock, the moment in which the subject and his/her experience of that world is exploded, becomes a moment for interrogators to exploit. And it becomes exploitable because of its connection to trauma. Trauma is here understood, via a then nascent psychoanalysis, to be a shattering of the self; a shattering that disturbs and undoes an otherwise normal, coherent, subject.

The means of exploiting the 'interval' are a range of 'non-coercive techniques'. In addition to advocating that interrogators manipulate the environment, the manual plays up the interpersonal dynamics of interrogation, identifying a number of persona that the interrogator can inhabit in order to shape the interaction with the interrogate and a number of techniques that they can use, in addition to controlling the environment. These are no less affective. One such technique for intensifying and capitalising on the 'moment' of shock is described in the manual as the 'Alice in Wonderland' technique. The aim is to confuse the subject and create a specific atmosphere:

> The confusion technique is designed not only to obliterate the familiar but to replace it with the weird. ... Sometimes two or more questions are asked simultaneously. Pitch, tone, and volume of the interrogators' voices are unrelated to the import of the questions. No pattern of questions and answers is permitted to develop, nor do the questions themselves relate logically to each other. In this strange atmosphere the subject finds that the pattern of speech and thought which he has learned to consider normal have been replaced by an eerie meaninglessness[11] (76).

Whilst not the focus here, it is worth noting that these and other techniques are not only found in the sites of intentional suffering that mark military interrogation. They are also present as part of a broader pedagogy of police interrogation that sees interrogation as an interpersonal dynamic staged with the aim of extracting information from one party in the exchange. What marks out Cold War-era CIA interrogation is the concurrent total control of the environment within which the interpersonal dynamic of interrogation occurs. The desired effect being the reduction of the subject to nothing but an effect of a form of total 'milieu control'. Milieu control comes to be enrolled into a relation of violence: where violence is understood in its most elemental as a force exerted by one thing over another. What the CIA manual also does is align the trauma of the interval of 'shock' to

11 These techniques are named; going next door, nobody loves you, the all seeing eye (or confession is good for the soul), the informer, news from home, the witness, joint suspects, Ivan is a dope, joint interrogators, language, the wolf in sheep's clothing, and Spinoza and Mortimer Snerd. The latter technique, for example, involves 'continued questioning about lofty topics that the source knows nothing about [that] may pave the way for the extraction of information at lower levels' (KUBARK 1963: 75). The interrogatee feels 'relief', or 'positively grateful', when they are asked a question they can answer (75).

the evocation of 'feelings of guilt'. The manual picks up and expresses Freudian theories of guilt (Leys 2007). Guilt is made into a normal feature of people's emotional lives. So the manual explains that: 'Most persons have areas of guilt in their emotional topographies, and an interrogator can often chart these areas just by noting refusals to follow certain lines of questioning' (KUBARK 1963: 66). Unlike in the first version of 'debility, dependency, dread', the subject is no longer solely defined by their responsiveness to the environment. Rather, the subject is governed environmentally *and* as an affective being with an 'emotional topography' that can be manipulated (as in guilt) or shattered (as in the trauma of the 'interval of suspended animation'). Two versions of affect come to coexist and blur with one another: environmental and psychoanalytical.

By becoming part of the apparatus of Cold War violence, 'debility, dependency, dread' has gone through a number of transformations as a version. It becomes a state to be induced in a potential enemy rather than a state to be counteracted in an actual friend; it becomes a state that can be 'inscribed in reality' through physical violence as well as environmental manipulation; it becomes known through a version of Freudian social psychology in addition to behaviouralism, information theory and cybernetics. Moreover, and perhaps most importantly, the relation with life that is produced through the apparatus is different. Brainwashing is a threat in the first apparatus and knowing 'debility, dependency, dread' is a way of counteracting that threat. The aim is not to stop 'debility, dependency, dread', but to produce subjects who can bear the experience of environmental control without disclosing information. What matters is not, then, the well-being of the subject. The relation with life is not providential. Instead, 'debility, dependency, dread' is named and known in order to produce a particular type of military subject and body: a subject that can retain information no matter what. Various US army and air force 'stress inoculation' programmes at the time of the Cold War attempted to produce a resistant subject who can dwell within an environment of 'total control' without disclosing information. What is protected is information. What is cared for are state and military secrets. By contrast, the aim when interrogating an enemy is the opposite: to produce subjects who cannot bear the environment, who cannot live through such conditions, and who will disclose information to their capturers and, ultimately, be neutralised as an enemy. Information is made to move, secrets are revealed. Undone by an environment that damages, the susceptible captor is subject to the will of the interrogator in an asymmetrical relation of dependency and violence.

3.5 Creating Unliveable Life

'Debility, dependency, dread' takes on a different existence as the Cold War apparatus of violence changes and new elements are integrated; Freudian versions of regression coexist with cybernetics, the CIA uses university psychologists, and the Cold War comes to be fought through an array of proxies. It is through these elements that 'debility, dependency, dread' becomes part of the specific apparatuses of violence that made up torture in the Cold War. As part of these apparatuses, 'debility, dependency, dread' is close to what Barad (2007: 149) terms a 'specific determinacy' with distinctive properties (involving distinctive claims about bodies and subjects based on specific versions of emotion), meaning (understood in the context of assumptions about the active and passive subject) and boundaries (in that 'debility, dependency, dread' is related to other affective states, including guilt and trauma but also an understanding of normal emotional topographies).

What 'debility, dependency, dread' is changes as part of the apparatus. Change is partly a consequence of the elaboration of new 'urgent needs', but it is also a function of how a 'thoroughly heterogeneous' ensemble comes together. There is a dynamism to ways of articulating and accounting for affective life. Let us illustrate this dynamism by returning, briefly, to the weaponisation of a song about a purple dinosaur in the detention centres that are part of the War on Terror. For what happens is that 'debility, dependency, dread' becomes indeterminate. Not because a named affect as an entity is taken to be multiple and is never stabilised as a coherent thing in itself, as we saw in the previous chapter's example of how morale emerged as an object-target. But because how 'debility, dependency, dread' had been propertied and bounded (to use Barad's (2007) terms) becomes indeterminate as a new apparatus forms: torture in spaces of detention that intensify a specific element of the Cold War – interrogation practices used by the US and its proxies in Latin America – in response to a specific enactment of the threat of terrorism, a new 'urgent need'. Still housed in the nominal unity of a name, 'debility, dependency, dread' comes to blur with the other named and many unnamed affects that make up the spaces and practices of torture. It also blurs with the conditions that enable torture to happen and extend the apparatus of violence beyond scenes and sites of torture.

'Futility music' is but one way in which 'debility, dependency, dread' was intentionally produced through interrogation practices in the War on Terror. As McCoy (2006: 8) details, other practices relied upon and were organised around 'simple, even banal procedures – isolation, standing, hot and cold, light and dark, noise and silence – for a systematic attack on all human senses'. Such 'simple, even banal procedures' have their origin in the behavioural experiments of Donald O. Hebb, and first find expression in the transnational and translocal networks of people, things and ideas that make up interrogation in the Cold War. Moving from the Cold War into and through the networks of military thought and multi-agency personnel[12] connecting Guantánamo Bay with rendition sites and the detention network of occupied Iraq, 'systematic psychological torture' became widespread throughout the prison complex in Iraq and probably Afghanistan

12 It is worth noting that most of the abuses in sites of detention appear to have been committed by multi-agency teams of CIA personnel and contractors, including 'Behavioral Science Consultants' (Cusick 2006; McCoy 2006).

(Borchelt 2005). Breaking with then US army and marine interrogation doctrine based on establishing 'rapport' or 'empathy' between interrogator and prisoner, practices of 'no-touch interrogation' were advocated and used that have clear roots in the KUBARK manual and repeat a type of violent 'environmentality'.[13] Inside containers, prisons and other spaces individual prisoners were 'conditioned' through techniques that travelled from Guantánamo Bay to Iraq and repeated in new arrangements the sensory deprivation and self-induced pain the KUBARK manual advocated. A US military intelligence memo summarises some of the 'techniques' used by military intelligence in Iraq, many of which have clear echoes of the KUBARK 1963 techniques coupled with a sovereign power that acts directly by marking and maiming:

1. Dietary manipulation – minimum bread and water, monitored by medics;
2. environmental manipulation – i.e. reducing A.C. [air conditioning] in summer, lower [ing] heat in winter;
3. sleep management – for 72-hour time period maximum, monitored by medics;
4. sensory deprivation – for 72-hour time period maximum, monitored by medics;
5. isolation – for longer than 30 days;
6. stress positions; and
7. presence of working dogs (US Military Intelligence: Interrogation and Counter-Resistance Policy, January 2004; cited in McCoy 2006: 139).

The emphasis on environmental control in Farber and colleagues' original response to the threat of Communist brainwashing can be found in these and other techniques that were used by coalition forces across various sites of internment and imprisonment during 2003. Targeting detainees as affective beings, the above techniques shape the environment in order to affect prisoners, as well as act on the body directly in a reminder of the 'policy of terror' that Foucault (1977: 49) shows

13 At the time, *Army Field Manual 34-52* drew strict lines between psychological coercion and psychological methods of interrogation, as did FBI interrogation policy and it seems practice. In the wake of the Abu Ghraib scandal this distinction has been reiterated in a subsequent iteration of interrogation doctrine (see *Army Field Manual on Interrogation* (2006): 5–21). It is worth noting that the prohibition in doctrine is often on pragmatic grounds – both in terms of protecting US military personnel from sanction under the Geneva Convention but also in terms of the usefulness of 'inhuman treatment'. So *Army Field Manual 34-52* stressed that 'Therefore the use of force is a poor technique, as it yields unreliable results, may damage subsequent collection efforts, and can induce the source to say whatever he think the interrogator wants to hear' (*Army Field Manual 34-52* cited in Borchelt 2005: 7). Whilst explicitly prohibiting a number of the techniques such as exposure to loud music or extreme temperatures, the 2006 iteration of the *Army Field Manual on Interrogation* does contain an Appendix (M) that legitimates certain abusive techniques under 'special' circumstances, including 'extreme isolation' and 'sleep manipulation'.

sovereign power can work through. A catastrophic form of power is exercised through the means of an environmental technology. In the War on Terror, conditioning meets sovereign power in hidden scenes of torture – scenes without the 'glory' Foucault attributed to sovereignty.

Every element of the environment of the prison was stipulated in an attempt at 'total' control that echoes the fantasies of control we find in 'total war'. Extending beyond determinate sites such as Abu Ghraib, the apparatus of violence includes elements from Cold War training manuals, whilst also involving new elements that make up the violent geographies of the so-called War on Terror. Not least the withdrawal and abandonment of the juridicial order in sites that became a 'strange hybrid' (Gregory 2007: 206) of military prison and ordinary prison and the excess of 'elaborate regulatory efforts' that Johns (2005: 614) has documented. Alongside the memoranda, directives and orders, of particular importance was the changing topology of friend/enemy that interrogation in Iraq was caught up in. Blurring legal and moral distinctions between friend and enemy, combatant and non-combatant, US military intelligence and an array of private contractors to whom interrogation was outsourced merged detention and interrogation policies in their treatment of 'security detainees' or 'security internees' (Gregory 2007: 221–2). These elements come together with other proximate causes: the brutality of the American prison complex in which some of those responsible for the worst abuses had worked; the chaos and disorder of post-occupation Iraq; a structure of feeling in which the threat of terror loomed over the here and now, amongst others.

In this apparatus of violence, 'debility, dependency, dread' is produced as an environmental mode of power converges with a sovereign mode of power across the improvised sites of detention that contain (potential) combatants. The most serious violations of international humanitarian law between March 2003 and November 2003 occurred in sites ran by military intelligence (including the military intelligence section at 'Abu Ghraib Correctional Facility'). Cataloguing the inhuman and degrading treatment afforded to persons deprived of their liberty, the 2004 International Committee of the Red Cross (ICRC) report into abuses was one of the first to bear witness to the violence of prison sites. By carefully detailing the routine use and reworking of the 'coercive' and 'non-coercive' interrogation techniques first found in the KUBARK manual during arrest, internment and interrogation, the report returns us to the cargo container in which a children's song of love was repeatedly played and played and played and played. The report also presents our final version of 'debility, dependency, dread' as an effect of an apparatus of violence that emerges in response to the threat of the terrorist or insurgent. Networked in form, shifting between presence and absence, the terrorist or insurgent threatens because it blurs and merges with the population. Following on, peace and war blur and merge, as do interrogation and imprisonment and arrest (Khalili 2012). The ICRC's report bears witness to the vortex of violence that is the result, detailing the most frequently alleged methods of ill-treatment during interrogation in a style now common to humanitarian witnessing. They include:

- Being held in solitary confinement combined with threats (to intern the individual indefinitely, to arrest other family members, to transfer the individual to Guantanamo), insufficient sleep, food or water deprivation, minimal access to showers (twice a week), denial of access to open air and prohibition of contacts with other persons deprived of their liberty;
- Being paraded naked outside cells in front of other persons deprived of their liberty, and guards, sometimes hooded or with women's underwear over the head;
- Acts of humiliation such as being made to stand naked against the wall of the cell with arms raised or with women's underwear over the head for prolonged periods – while being laughed at by guards, including female guards, and sometimes photographed in this position (International Committee of the Red Cross (2004) in Danner 2004).

Across the detention sites of the War on Terror, US military intelligence interrogators aimed to create a particular type of environment that takes us back to the unnamed subject of the initial 1950s experiments and the attempt to 'condition' his reactions. Allied to the techniques that first weaponised behaviouralism by merging it with a version of regression, were acts and techniques of sexual torture that drew on orientalist knowledge of a monolithic Muslim 'culture' that circulated in US policy, military and academic networks. Puar (2007) stresses the importance of Raphael Patai's *The Arab Mind* in developing techniques that depended on inducing shame. Puar describes how an apparatus of violence congeals to produce the Muslim body, rather than the abstract body-and-environment, as an object of torture in the occupations of Iraq and Afghanistan:

> This kind of torture directed at the supposed "Muslim Terrorist" is subject to the normativizing knowledges of modernity that mark him (or her) both as sexually conservative, modest and fearful of nudity (and it is interesting how this conceptualization is rendered both sympathetically and as a problem), as well as queer, animalistic, barbarian, and unable to control his (or her) urges (Puar 2007: 86).

Normativising knowledges of shame and sexual repression blur with behavioural and cybernetic knowledges of the individual-in-environment in a vortex of humiliation. The object-target of incarceration is racialised and sexualised. 'Debility, dependency, dread' blurs into other named affective states as new elements are added to the apparatus of violence in the War on Terror: the Muslim subject is stripped of everything that makes him or her a subject, but this is done through the production of shame alongside an environmental conditioning. A version of repression comes to replace the regression that the KUBARK manual

(1963) was organised around, whilst emphasis on creating 'psychological shock' remains. Public humiliation designed to induce shame comes to coexist with what Leys (2007) terms the 'identificatory logic of guilt' that featured in the 1963 manual and resulted in techniques of interrogation that aimed to 'intensify' 'guilt feelings'. Whilst some aspects of counter-insurgency doctrine warn against the production of shame,[14] in spaces of incarceration shame was created through scenes of public humiliation. 'Debility, dependency, dread' moves from being a distinct entity, to being one element that mixes with other affects of violence in a way that exceeds the ordered form of the humanitarian reason exemplified by the bullet-pointed list of the Red Cross. In this process 'debility, dependency, dread' is no longer a distinct entity. It blurs with other versions of what affect is and does, including shame and guilt. 'Debility, dependency, dread' becomes indistinct amid the horror of torture and uncertainty of open-ended incarceration.

3.6 The Necessity but Insufficiency of Critique

> Critique only exists in relation with something other than itself (Foucault 1996: 383).

In a short piece on Foucault's concept of apparatus, Agamben (2009: 13) links the term to what he calls a '[g]eneral and massive partitioning of beings into two large groups or classes'. An apparatus, for Agamben, is a generic term for literally anything that has the capacity to manipulate living beings (he links the term to Heidegger's *Ge-stell* as well as *oikonomia* and *dispositio*). Thus:

> I shall call an apparatus literally anything that has in some way the capacity to capture, orient, determine, intercept, model, control, or secure the gestures, behaviors, opinions, or discourses of living beings (Agamben 2009: 14).

In Agamben's development of the term apparatus, for at this point he has moved beyond an interpretation of Foucault, apparatuses and living beings are locked in a 'relentless fight' (14). As Legg (2010) makes clear in an insightful and careful reading of the piece, 'living beings' is close to Agamben's formulation of *zoë*, the simple fact of living that requires qualification to become *bios*. It is this distinction between apparatuses and living beings that has been in the background of this and

---

14 Some critiques of the US interrogation policy have focused on its so-called 'blowback' effect: inducing shame in the male population begets anger that begets an increase in insurgency, or so the chain of reasoning goes (a similar argument has recently been used against drone strikes). Whilst perhaps useful as an argument against torture, it uses a pragmatic-speculative calculus of usefulness as its criteria for whether torture should be used, a move that allows the counter-position that torture is useful rather than an ethical or moral ground that torture is always wrong.

the previous chapter and I have attempted to rework and think outside of. Rather than counterpoise affective life to ways of knowing/representing affective life, my emphasis has been on what we could call the life of ways of mediating affective life. Through the examples of morale and 'debility, dependency, dread', we have seen how knowing, rendering actionable and intervening always involves versions of what affect is and does, alongside the determination of specific entities. Over the course of the two chapters I have touched on some of these different versions of affect: behaviouralism, cybernetics, neuroscience, crowd psychology, and so on. There are many others. What this means is that it is not enough to counterpoise the force or powers of affective life against apparatuses that mediate that life but are, supposedly, single, stable and coherent. Apparatuses are lively. They may themselves be non-coherent and subject to change; sometimes through the redeployment of existing elements in new arrangements, sometimes through the introduction of a new element, sometimes through moments of intensification, sometimes through the loss of an existing element, and sometimes through encounters with other apparatuses.

For example, changes in 'debility, dependency, dread' relate to a shift from a combination of behaviouralism and cybernetics to a popularised version of Freudian regression and cross-cultural social psychology. In the sense that Barad (2007) uses the terms, 'debility, dependency, dread' is propertied and bounded; at once both a representation of an experience that is used to make sense of that experience and a new element in social life through which bodies are made and remade. Outside of the temporary integration of a differential field, 'debility, dependency, dread' is indeterminate. Within the context of that integration, it is rendered measurable and actionable. In this sense, the apparatus of military violence that provides the context for 'debility, dependency, dread' is akin to a dynamic field of possibilities that conditions how affective life can be rendered actionable. It follows that 'debility, dependency, dread' is not a single thing. It changes as apparatuses of experimentation change and it changes as it moves between apparatuses that come to be organised around detention camps and internment centres. 'Debility, dependency, dread' is something different in the apparatus of late 1950s behaviouralism and the response to Cold War anxieties than it is in the sites of intentional suffering that make up Abu Ghraib, Guantánamo Bay and the other known and unknown sites that compose the War on Terror. Also, 'debility, dependency, dread' does not always remain a determinate entity with stable properties, meaning and boundaries. It shifts between being rendered determinate in experimentation or during procedures and becoming indeterminate as it comes to mix with and become inseparable from other object-targets. For example, images of the corporeal violence of torture come to life in some of the mediatised humanitarian apparatuses that enable the writing of this chapter and change the meaning of 'debility, dependency, dread': an apparatus of humanitarian witnessing that makes present harm and damage as an ethical response to suffering, and an apparatus of the circulation and distribution of digital images of torture through which moral publics are constituted around the condemnation of torture.

Differences in how 'debility, dependency, dread' is propertied and bounded are not due to some form of meta shift in a society's emotions – say from a culture organised around guilt to shame (compare with Leys 2007). Rather, differences are a function of changes to the relations between the discursive and non-discursive elements that make up apparatuses, the 'urgent needs' that animate apparatuses, the way in which specific intentions are layered into apparatuses, and the institutions, procedures and other mechanisms that make up apparatuses. Whilst not necessarily always the case, this means that a named affect might be multiple. We have seen in this chapter and Chapter 2 how morale and 'debility, dependency, dread' are different entities depending on the apparatus through which they are articulated and on the partial connections between those apparatuses. Differences in what we might name as a single affect will be a function of the specific composition of apparatuses. In this sense, attempts to represent and intervene may themselves have an affective life. Just as hope became attached to targeting morale in the context of the total mobilisation of life in 'total war', so a hope became attached to 'debility, dependency, dread'. Initially, 'debility, dependency, dread' is inseparable from the hope of protection from a threatening USSR. In the War on Terror, it becomes enrolled in the hope of defeating a spectral networked insurgency that threatens because of its absence, and the promise of protection to a population supposedly threatened by the unpredictable emergence of insurgencies (Khalili 2012).

Based on the previous two chapters we can summarise the vocabulary proper to the first of the three processes of mediation the book examines: affective life as an object of knowledge, a target of intervention and a means of intervening in life. What I have offered in the first two chapters is the beginnings of a method that traces how specific entities (such as 'debility, dependency, dread' and morale) are constituted through apparatuses that involve distinct versions of what affect is and does. Those apparatuses are multiple, inscribed in relations of power/knowledge, and produce specific entities with distinct properties, boundaries and meaning. Whilst the emphasis in the first two chapters has been on affective states that are named in military violence, the approach to following the determination of distinct versions holds for other ways we have of accounting for and articulating affective life. Turning back to the examples of retail atmospherics at the start of the previous chapter, now in comparison with morale and 'debility, dependency, dread', we can see how ScentAir involves specific versions of affect – that make affect into a biological and neurological type of thing – and materialise specific affective entities – ways of articulating and accounting for consumer desires in a manner that attempts to grasp and render actionable the unpredictability of consumer desire, motivation and action by explicating and attempting to shape the affective background of sites.

Understanding how affective life is represented and intervened in necessitates a particular practice of critique, despite the recent arguments against critique in a range of work inspired by the 'affective turn'.[15] In a short interview conducted after the 1981 election of François Mitterrand and first published in May of that year, Foucault describes the type of critique I have in mind. For Foucault, critique is a certain type of activity that is indispensible for any transformation:

> Criticism is a matter of flushing out that thought and trying to change it: to show that things are not as self-evident as one believed, to see that what is accepted as self-evident will no longer be accepted as such. Practicing criticism is a matter of making facile gestures difficult (Foucault 1988: 155).

Here critique is, on the one hand, a practical activity and, on the other hand, a permanent disruption that must agitate, unsettle and ultimately break with modes of thought. As such, critique becomes a way of agitating the 'free atmosphere' in which transformation takes place. Critique becomes an atmospherics based on an attempt to bring a seemingly settled apparatuses to crisis. Later on in the interview, Foucault makes clear how a specific tone animates a practice that is always orientated to an open future it cannot pre-programme. Foucault is asked about his optimism:

> There's an optimism that consists in saying that things couldn't be better. My optimism would consist rather in saying that so many things can be changed, fragile as they are, bound up more with circumstances than necessities, more arbitrary than self-evident, more a matter of complex, but temporary, historical circumstances than with inevitable anthropological constants (Foucault 1988: 156).

It is this form of optimism that is needed when confronted by apparatuses. Not the defeatism of offering yet another account of the domination of life without limit or remainder, but a mode of engagement that discloses the fragility of how a multiplicity of bits and pieces are integrated through a strategic relation. How, then, might a critique that aimed to produce such 'turning points' proceed in relation to those apparatuses that mediate affect as an object-target? What questions might such a critique ask of apparatuses in order to disclose their fragilities and

15 Massumi (2002a: 12–13) calls for a 'shift to affirmative methods: techniques that embrace their own inventiveness and are not afraid to own up to the fact that they add (if so meagrely) to reality'. See Chapter 4 for a brief discussion of Massumi and Sedgwick's engagement with critique. Latour (2004a), for example, argues that critique has 'run out of steam' on the grounds that it is based on a practice that stands apart from the world and then stands in judgment over it, forever telling stories that give a causal role to the same actors (globalisation, capitalism, neo-liberalism, and so on). The result, for Latour, is that critical social science has come to have much in common with conspiracy theorists: 'it's the same appeal to powerful agents hidden in the dark acting always consistently and continuously, relentlessly' (Latour 2004a: 229). For me, Latour's description of critique is too one-dimensional and does not pay enough attention to the actual practice or attitude of critique or the multiplicity of tones that critique involves – hence my discussion of Foucault on critique.

understand them as active mediations? I think there are three that come together to form a mode of engagement with affective life.

First, how do discursive and non-discursive elements come together and hold together, if only ever temporarily and if only with continual work? In the example of morale in the previous chapter we saw that morale became an object-target through a kind of practical military psychology, bombing reports, techniques of measurement and much more. What is important to trace is whether and how knowledges of affective life are inscribed in specific power relation or sets of power relations. Second, how do the elements in an apparatus change, whether through redeployments or intensifications and how does this occur in relation to an 'urgent need'? Apparatuses are more than just a stable or formed order. They are strategic. In the example of morale the 'urgent need' was to sustain war by harnessing the energies of a population. 'Debility, dependency, dread' was born from a need to ensure a balance of powers amid the heat of the Cold War. But that 'urgent need' or 'demand' has itself a life, it emerges and changes. Third, how does the apparatus express and enact a specific topology of power and in what ways does this involve a cut within life between a valued and devalued life? In the two cases we saw different relations of power that cannot be understood by relying on a model that assumes the manipulation of affective life by a centred and powerful subject. Morale was targeted by governments in a combination of vital power and sovereign power. 'Debility, dependency, dread' was produced through a type of violent sovereign power aligned to an environmentality. In both cases we also see a cut is made within life between lives that are valued and lives that are devalued. When attempting to secure morale, specific categories of raced and classed peoples are identified as causes of 'bad morale'. In the example of 'debility, dependency, dread' in the War on Terror, Muslim bodies that supposedly have the potential to threaten are damaged and destroyed through racialising apparatuses. A key question is, then, how attempts to represent and intervene in affective life become part of the (re)production of harms, sufferings and damages.

Critique is a specific ethos of engagement with apparatuses, one that aims to bring an apparatus to crisis, and one that is imbued with the type of optimism Foucault spoke of in 1981. This leads to a conclusion from this and the last chapter: understanding how affective life becomes an object-target for forms of power is a precondition for developing affirmative relations with affective life. However, critique is necessary but not sufficient. It is necessary not only because it shows the fragility of things, but also because it describes specific relations of power and the precise operation of the apparatuses through which affective life is mediated. It allows analysis to move beyond general claims about epochal change and affect. Yet, critique alone is insufficient because affective life is mediated and organised in ways that exceed the making of affective life into power's object or the object of power. This is the topic of the next chapter of the book where I turn to the concept of affect directly and chapters 5 and 6 that describe how collective conditions mediate life affectively. We have seen hints of other ways of thinking about affective life in the background to the stories I have told about morale and 'debility, dependency, dread'. On the one hand, I have hinted at how various collective affects may pressure and limit the problems that apparatuses relate to. Note, for example, the importance of the structure of feeling named as Cold War fear to the invention of 'debility, dependency, dread' as a way of countering Communist 'brainwashing', or perhaps the structure of feeling of emergency that formed the background to interrogation policy in the so-called War on Terror, to name but two. On the other hand, and in the background to the two examples, has been a sense that affective life as it is lived exceeds apparatuses that attempt to know and name affects. Ordinary affective life may be mediated by specific apparatuses, but it is not reducible to those apparatuses and may involve other forms of organisation and processes of mediation.

3.7 Afterword: A Dream

The multiple relations with life that exceed critique and are necessary to understand the mediation of affective life were evoked by an anonymous philosopher in an interview in 1980. The words have stayed with me since I first read them as a PhD student. For that reason, I present the dream of a type of engagement with the world that reverberates across the rest of the book as I turn to discuss other processes, forms and effects of mediation:

> I can't help but dream about a kind of criticism that would try not to judge but to bring an oeuvre, a book, a sentence, an idea to life; it would light fires, watch the grass grow, listen to the wind, and catch the sea foam in the breeze and scatter it. It would multiply not judgments but signs of existence; it would summon them, drag them from their sleep. Perhaps it would invent them sometimes – all the better. All the better.[16]

What would it mean to 'bring to life' and multiply and invent signs of existence? What is the relation between such an ethos and the style of critique advocated here? The following chapters engage with these questions in attempting to outline an analytics of affect adequate to the complexities of affective life. What I take from Foucault's cryptic, hopeful, dream is the necessity of description and speculation. An ethos of engagement with the world that begins by affirming the real effects that strange things like affects, structures of feeling and affective atmospheres can and do have. My emphasis in chapters 1 and 2 on versions of affect has one further implication: what follows is but one version of what affective life is and does that coexists, harmoniously or not, in a crowded field with multiple other versions. It is a version that starts not from a transhistorical constant – say a limited number of basic emotions – but from the contingency of how bodies come together in encounter.

16 Michel Foucault (1997) in conversation with Christian Delacampagne. First published in *Le Monde* 6–7 April 1980.

Chapter 4
The Imbrication of Affect

> Nothing can be reduced to anything else, nothing can be deduced from anything else, everything may be allied to everything else
>
> (Latour 1988: 163).

4.1 Everything is Affective

A lost hope for recognition by another person; a persistent hope that something might change; the intrigue and indifference that might surround a rumour; the disgust and interest that attach to torture; the awe and trembling as bombs fall; the isolation and despair of incarceration; love and soothing words as my daughter cries and I pause before trying to write; exuberant trades as markets rise, confident speech; a hope through a felt past…

Any consideration of affective life begins with an affirmation: that affects are real forces that are part of the composition of common worlds rather than mere epiphenomena that can be reduced to the operation of apparatuses. Affects of hope, intrigue, exuberance, and so on, are as real as the infrastructures, classes, Gods, and other social factors and forces that populate life. This claim is hardly a novel one. It resonates with a subterranean current of attempts to describe societies and cultures in terms of their affective life that predates the recent 'affective turn' in cultural theory (Blackman 2012; Seigworth and Gregg 2010). For one example, consider the economic writings of Gabriel Tarde. In his economic psychology, Tarde describes the economy as an organisation of passions, a type of (im)personal affective complex. Affectively imbued encounters between economic actors form a sort of ever-present/never-present affective 'background' that both enables and is reworked through economic apparatuses:

> From salesman to client, from client to salesman, from consumer to consumer and from producer to producer, whether competing or not, there is a continuous and invisible transmission of feelings – an exchange of persuasions and excitements through conversations, through newspapers, through example – which precedes commercial exchanges, often making them possible, and which always helps to set their conditions (Tarde, cited in Latour and Lépinay 2009: 39).

Tarde is not alone in his concern with how life and living are affective. From reflections on the panic of crowds in early twentieth-century crowd and mass psychology (Orr 2006), through to early Freudian work on transference in hypnosis (Green 1977), there have long been speculations on the transpersonal dynamics of affect. Nevertheless, it has been far easier to reduce the affective life of societies to the emotions of individuals, anchoring and enclosing feeling in an individual. What Tarde and the other writers in his wake invite us to do is consider life and living as everywhere affective, and to wonder about how affects are thereafter organised. On this account affect is not only an object-target for various apparatuses, no matter how indeterminate or changeable. In the background to the previous two chapters has been a second way of thinking about and encountering affective life and its specific and varied modes of mediation and organisation: that bodily capacities are irreducible to apparatuses. So, for example, a rumour designed to undermine morale may be greeted with laughter, hope might be kept alive even as 'debility, dependency dread' is produced, revulsion and captivation may mix as digital images of torture are viewed, or the introduction of a new atmospheric materiality may be met with a vague sense of discomfort and mild puzzlement. Bodily capacities are not linear effects of apparatuses and the acts, ideas and intentions that make them. To paraphrase Latour (1988: 163), affects cannot be reduced to anything else, affects cannot be deduced from anything else, but affects may be connected to anything.

Whilst insisting on the irreducibility of affects to something else, we must also be extremely cautious about any equation between affect and the vital powers of life, where those powers would always exceed the apparatuses that shape affective life. Once we understand affects as forces of existing, then everything may be affective. As Shaviro (2010) puts it, affect is an entirely generic notion. There is nothing devoid of affect. Massumi (2002a: 37) emphasises something similar, albeit in a different context, when he stresses how 'Affect, like thought or reflection, could be extended to any and every level, providing that the uniqueness of its functioning on that level is taken into account'. This means that affect is as neutral as a term such as meaning: affects such as hope, intrigue, indifference, and so on, are as linked to organisation as to contingency, to the systematic as to the disruptive, to structure and context as to the creative powers of events and the surprise of life.

My aim in this chapter is, then, to offer an infralanguage proper to affect as a bodily capacity that focuses attention on how a body's 'charge of affect' is mediated and organised. I consider how we might think affect as a bodily capacity in relation to the account of apparatuses in the preceding two chapters and the description of collective affects in the following two chapters. For that reason, this chapter is a hinge. I discuss the second of the three ways in which affective life is mediated and organised: through the assembling of the encounters from which affects emerge and become expressed. And I do so by focusing on just one of the now multiple translations of the term affect: affect as 'capacities to affect and be affected' as developed through Gilles Deleuze's (with Felix Guattari) encounter with Baruch Spinoza, and subsequent experiments by Brian Massumi and others in contemporary cultural theory. As I set out in Chapter 1, bodily capacities are two-sided *capacities* to affect and to be affected. On the one hand, affect refers to a capacity to be affected through some form of affection. On the other hand, affect describes the capacity a body can have to affect something outside of itself. 'Being affected-affecting' are therefore two sides of the same dynamic shift, or change, in what a body is and does since 'when you affect something, you are at the same time opening yourself up to being affected in turn' (Massumi 2002c: 212). It is this relational version of affect as a 'force of existing' (Deleuze 1978) that, in geography at least, has become associated with a range of non-representational theories (see McCormack 2003; Bissell 2008; Dewsbury 2000; Ash 2010; Simpson 2008; Lim 2010; Swanton 2010; Lea 2008). My wager is that this approach provides resources for attending to life in all its richness without grounding real experience in bodily physiology or brute physicality, or an already constituted, naturalised, subject who expresses emotions. Let us turn to set out some of the starting points for this approach through examples of two encounters before exploring their implications for understanding how affective life happens. Starting points that make the term affect into a kind of sensitising device; a way of disclosing life that orientates inquiry to how multiple forms of mediation come together in encounters.

4.2 'Capacities to Affect and be Affected' and Emotions

> [I]n the street I run into Pierre, for whom I feel hostility, I pass by and say hello to Pierre, or perhaps I am afraid of him, and then I suddenly see Paul who is very very charming, and I say hello to Paul reassuredly and contentedly. Well. What is it? In part, succession of two ideas, the idea of Pierre and the idea of Paul; but there is something else: a variation also operates in me – on this point, Spinoza's words are very precise and I cite them: (variation) of my force of existing, or another word he employs as a synonym: vis existendi, the force of existing, or potentia agendi, the power [puissance] of acting, and these variations are perpetual (Deleuze 1978, no pagination).

In his 24 January 1978 lecture on idea and affect in Spinoza, Deleuze offers what he describes as a 'stupid' example of an encounter. Two people pass one another, they say hello, before one of the two then sees and says hello to another person. Deleuze ends his description of the encounter with a definition of affect – 'a variation also operates in me' – that accompanies and is inseparable from the emotions (for example, 'hostility') that are layered into the encounters with Paul and Pierre. Through the example we reach a definition of affect: bodily capacities, or what Deleuze in the lecture refers to as 'force of existing' or 'power of acting', after he first defines affect generally as 'any mode of thought insofar as it is non-representational' (1978, no pagination)[1]. Examples might include the

1 Deleuze (1978) goes on to describe a variation in the force of existing: 'When the idea of Paul succeeds the idea of Pierre, it is agreeable to say that my force of existing or my power of acting is increased or improved; when, on the contrary, the situation is reversed, when after having seen someone who made me joyful I then see someone who makes me sad, I say that my power of acting is inhibited or obstructed'. Here we come to the two

blush of a body shamed (Probyn 2000b), the heat of a body angered (Katz 1999) or the restless visceral tension of a body bored (Anderson 2004b). Feelings are expressions of a body's 'charge of affect' and, as such, always both a reflection of how encounters happen and enacted in encounters. Feelings imply the presence of an affecting body. But variations do not simply happen to a blank body in space or in time. Feelings express the affected body's existing capacity to affect and be affected, or what Massumi (2002a: 15) describes as perpetual bodily changes 'in which powers to affect and be affected are addressable by a next event and how readily addressable they are'.

By starting from variations in bodily capacities, neither Deleuze nor Massumi reduce affect to a physiological mechanism or a subjective experience (Brown and Stenner 2009: 131). Rather than describe the minutiae of non or not yet conscious material perception, thinking in terms of a 'force of existing' directs us to what a body can do.[2] And how a body can affect and be affected continuously varies as encounters happen and capacities emerge, change and are realised. We find this attention to the dynamism of 'forces of existing' throughout Deleuze's early writings on Spinoza, his unsystematic systematisation of affect in the collaborations with Guattari and Parnet, and his encounters with art, literature, music and cinema: the horse as seen by little Hans (Deleuze and Guattari 1987: 284), the 'point' of Jack the Ripper's knife (Deleuze 1986: 97), the 'violent affect' of love between Heathcliff and Catherine that Emily Bronte invents (Deleuze and Guattari 1994: 175), a 'disreputable' man in a Dickens novel who is found as he lies dying (Deleuze 2001: 28).

Deleuze's example is a simple one. We know nothing about Pierre or Paul, the reasons for 'being afraid', or the street in which the encounters happen. Nevertheless, the example reminds us to avoid the mistake of thinking that a body's 'charge of affect' is best thought of as a natural, or at least under-socialised, phenomenon that is somehow equivalent to unmediated, biological processes. For the change in capacity to affect is mediated by, amongst other things, the reasons for the narrator's 'hostility' or their 'being afraid'. It might also be mediated by the setting of the street, including the street's atmosphere(s) (see Chapter 6).[3]

---

poles on the 'melodic line of continuous variation' – joy and sadness. Given their rigorous use it is important to note that they do not necessarily function on the conventional axes of happy–sad or pleasant–unpleasant, nor refer to a state that is possessed by a subject. Sadness is, rather, a name given to the affect insofar as it involves a diminution in the power of acting, whereas joy is a name given to the affect insofar as it involves an increase in the power of acting. In other words joy and sadness are two names given to limits and thus any determinate affect, any specific line of continuous variation, is likely to be a mixture of the two.

2 The emphasis on what a body can do resonates with Latour's (2004b) definition of the body as what learns to be affected and his substitution of the question 'What is a body' for 'what can a body do?'.

3 Ahmed (2004: 39), for example, argues that distinguishing between modalities can risk equating affect with immediate corporeal sensations and can thus create a 'distinction

Perhaps it is also mediated by the contented tone of the uttered greeting 'hello', the narrator's thoughts of whatever he or she was on their way to do, or the differences between Pierre and Paul. Let us turn to a second example – this time on a train, during a journey – to learn a little more about how bodily capacities emerge in encounters and how encounters are mediated by forces that exceed their spatio-temporal location.[4] For, as Clough (2007: 3) puts it, beginning from affect allows us 'to grasp the changes that constitute the social and to explore them as changes in ourselves, circulating through our bodies, our subjectivities, yet irreducible to the individual, the personal and the psychological':

> "Look, a Negro!" It was an external stimulus that flicked over me as a I passed by. I made a tight smile.
>
> "Look, a Negro!" It was true. It amused me.
>
> "Look, a Negro!" The circle was drawing a bit tighter. I made no secret of my amusement.
>
> "Mama, see the Negro! I'm frightened!" Frightened! Frightened! Now they were beginning to be afraid of me. I made up my mind to laugh myself to tears, but laughter had become impossible (Fanon 1986: 111–12).

The encounter is described by Frantz Fanon (1986) as part of his account of how processes of racialisation work through and accumulate in living bodies, specifically Fanon's emphasis on the psychic and somatic misery that colonialism inflicts. Unlike in Deleuze's 'simple' example, the encounter is mediated, predominantly by the racial formation of colonialism refracted through forces of sexual difference. Fanon describes a series of colonial affects that happen as the encounter happens; 'a tight smile', 'amusement that was not hidden' and the 'impossibility of laughter'. The encounter *constricts*: 'The circle was drawing a bit tighter'. Fear moves between Fanon and the child, amplifies and changes in the time of the encounter: 'Now they were beginning to be afraid of me. I made up my mind to laugh myself to tears, but laughter had become impossible' (Fanon 1986: 112).

---

between conscious recognition and "direct" feeling, which negates how what is not consciously experienced may still be mediated by past experiences'. This appears to only work as a critique if one assumes that affect is being used a) as equivalent to immediate, corporeal sensations and b) that there is not simply a pragmatic-contextual distinction between affect and other modalities but an absolute divide.

4 First published in 1952 and translated into English in 1967, Fanon (1986) is interested in the existential ambiguities of racialised subjects in the context of colonial relations between racist and racialised. As Chow (1999) shows, central to Fanon's complex portrayal of postcolonial subjectivity is the 'infernal circle' (Fanon, 1986: 116) of shame and longing-for recognition that Fanon presents as central to the (post)colonial subject formation of black men.

I juxtapose Fanon and Deleuze because, despite their many differences, their examples invite us to map how a body's 'force of existing' forms amid the ongoing (re)composition of encounters. There have been other examples of encounters in the book so far: the encounter with a plane that threatens to bomb, the encounter with a song about a purple dinosaur. Life is made up of innumerable encounters across and between different types of bodies. A body's 'charge of affect' enacts, reflects and expresses the agencement of encounters with other times and spaces. Encounters are made through repetitions. Something of the past persists in an encounter, any encounter contains reference to past encounters, and encounters are made through accumulated relations, dispositions and habits. Encounters also involve differences, in that as bodies come together in encounters life is opened up to what is not yet determined or is to be determined.[5] As patterns of difference and repetition, what happens in an encounter is never completely foreclosed. For this reason Spinoza's (1910) (Book III, Proposition 2, Scholium, 280) affirmation of openness has become so important to work on affect, providing an injunction to attune to change: 'nobody as yet has determined the limits of the body's capabilities: that is, nobody as yet has learned from experience what the body can and cannot do'. Faithful to Spinoza's affirmation, Massumi (2002c: 212) stresses that affect, therefore, provides 'a way of talking about that margin of manoeuvrability' in a situation, the 'freedom' that follows from asking '"where we might be able to go and what we might be able to do" in every present situation'. On this understanding, critique is no longer the default mode of engagement with affective life. Instead, techniques of thought and action must be invented to sense and perhaps cultivate the potential for change that may open up in the gap between affect and action. Conceived of as force, affect comes to name the aleatory dynamics of how encounters exceed the forces and processes that mediate them.

The emphasis on the autonomy of affect raises questions that I will return to later in the chapter, in particular through an example of several hopeful moments: is affect always in excess of determinations and must an attention to affect's excessive qualities be grounded in an ontological claim about the dynamism, exuberance and creativity of 'affect itself'? As Fanon's encounter shows, and Massumi does acknowledge, any 'margin of manoeuvrability' is always mediated; the encounter with a child on a train is mediated by the historical forces of colonialism and various racial schema that extend beyond the site of the encounter. Fanon (1986: 112) 'existed triply' in the encounter: 'I was responsible at the same time for my body, for my race, for my ancestors'. He stresses the curtailment of possibilities and shattered hope. His encounter constricts; reducing him to the black body that frightens and terrifies a child. I will return to the question of how encounters are mediated below. Before doing so, it is important to note that integral to this particular version of affect has been an analytic distinction between affect and other modalities, principally emotion/feeling. What the distinction affords, what it makes possible

5 The terms 'to be determined' and 'not yet determined' are taken from Shaviro (2009: 63).

to attune to, is how affective life is expressed in ways that scramble distinctions between the personal and impersonal. This is as long as we treat the distinction as a pragmatic-contextual one that helps us make sense of the complexity of affective life rather than an ontological claim that devalues the category of emotion and affirms only the escape of affect over any and all determinations.

Various processes of qualification translate the emergence of affect, and the expression of feeling, into emotions linked to subjectivities and identities (see Lupton 1998). In what is now a well-cited passage, Brian Massumi provides the clearest expression of how *from within an analytic distinction* affect and emotion pertain to different orders, with the former being closely associated with the contingency of social life and the latter being a secondary reduction of the 'openness' of affect.[6]

> An emotion is a subjective content, the sociolinguistic fixing of the quality of an experience which is from that point onward defined as personal. Emotion is qualified intensity, the conventional, consensual point of insertion of intensity into semantically and semiotically formed progressions, into narrativizable action-reaction circuits, into function and meaning. It is intensity owned and recognised (Massumi 2002a: 28).

I have already questioned this association of affect with openness, and argued that bodily capacities are always-already mediated, including by signifying apparatuses. For now, we should note that instead of affect and emotion existing in separate unconnected levels, we can stress multiple relations between emotion as a 'subjective content' and affect as bodily capacities. Emotion as qualified personal content feeds back into the emergence and organisation of affect, whilst at the same time being the most intense expression of the capture of affect *and* of affect's ongoing escape (Massumi 2002a: 35). Depending on the apparatuses through which emotions are qualified, some of these processes of qualification may involve the many ways in which some subjects may recognise, name, understand and reflect on emotions. Massumi (2002a), reworking Deleuze and Guattari's (1987: 65) pejorative comments on Cartesian passion as a pathology of subjectivist thought, stresses that an emotion is:

> the most intense (most contracted) expression of that *capture* – and of the fact that something has always and again escaped. … That is why all emotion is more or

6 Variants of this distinction have been taken up and repeated by a range of work in the affective turn, including but going beyond non-representational theory in geography. For example, Hardt and Negri (2004: 108) utilise a problematic distinction between emotions as a form of 'mental phenomena' whilst affects 'refer equally to body and mind'. Massumi's account is clearly more complex than this, given that it stresses the interimplication of affect and emotion and the way in which the 'insertion' of affect into signifying apparatuses results in variations in a body's 'force of existing'.

> less disorientating, and why it is classically described as being outside of oneself, at the very point at which one is most intimately and unshareably in contact with oneself and one's vitality (Massumi 2002a: 35, emphasis in original).

We might question whether emotion is always 'disorientating', but the result is that the literature on affect repeats the classical theme that emotion *is* the openness of a subject to its own self-difference in experience (Terada 2003). In breaking with the persistent link between emotion and the expression of a subject (Terada 2003), this account of emotion shares much with a range of recent work that has been based on the minimal specification that what are referred to as emotions are relational. For emotion may open up the subject to what might become a 'non-subjective experience in the form of self-difference within cognition' (Terada 2003: 3). What we also get a sense of here is the form of connection between affect and emotion (even if affect is given 'ontological priority' and emotion is reduced to a secondary capture). Instead of distinct levels, with no possibility of communication between them, processes of expression and qualification are part of the encounters in which bodily capacities are formed. From the perspective of understanding how bodily capacities emerge, a pragmatic-contextual distinction between distinct modalities allows us to trace how different processes of expression and qualification are part of encounters, and follow how they interact with other elements as bodies come together. This means that in practice affect, feeling and emotion are indistinguishable. It is not that there are, first, separate layers that are then, secondarily, connected. Rather, affects and emotions are always-already entangled with one another in encounters – encounters that mix and render indistinguishable the personal and impersonal (but may involve processes whereby emotions come to be felt as personal).

4.3 How is Affect Non-Representational?

This is, however, only one version of affect[7]. Other versions of affect come freighted with different assumptions about the ordering of social life and what

7 We could compare with Eve Kosofsky Sedgwick's (2003) emphasis on 'finitely-many' affects. Within the 'finitely-many' affects there are infinite graduations and, thus, infinite permutations as affects mix in encounters (Sedgwick and Frank 1995: 15). Sedgwick's starting point is the freedom of the affect system with regard to its aims, ends, time and object in comparison to the drives (which for Sedgwick after Silvan Tomkins (1995) are more constrained with regard to means-ends, the exception being sexuality). Refusing the 'binary homogenisation' or 'infinitizing trivialization' that discussions of difference are too often captured by, a claim about biology is what in part allows for Sedgwick's attention to the 'combinatorial complexity' of the affects. Sedgwick's account is rooted in Tomkins's (1995) identification of eight basic affects/emotions. The eight affects/emotions identified at the time of publication of volume one of *Affect, Imagery, Consciousness* were: interest-excitement (eyebrows down, track, look, listen), enjoyment-joy (smile, lips widened up and out), surprise-startle (eyebrows up, eye blink), distress-anguish (cry, arched

a body is and does. It is worth pausing and juxtaposing the emphasis on affect as bodily capacities with other prominent versions of affect offered in recent social and cultural theory. Whilst superficially similar, there are some important differences within what has been labelled as an 'affective turn' on the vexed question of how exactly affect is non-representational. By way of Deleuze and Fanon, I have offered one answer: affects are augmentations or diminutions of a body's 'force of existing' that are expressed in feelings and qualified in emotions (and where emotions/affects become indistinguishable in experience). What this version of affect invites us to pay attention to is encounters, and how capacities are made through processes of qualification and expression. But it also allows us to understand the geo-historicity of 'forces of existing' by tracing how processes of mediation become part of encounters. This means that affect is mediated in the sense that it is shaped by the participants in an encounter, rather than being exclusively organised through some form of representational-referential system of signification (on which, see section 5.2). Indeed, a body's 'charge of affect' is a function of both a series of immediate encounters and the geo-historicity of the body – the manner in which capacities have been formed through past encounters that repeat, with variation, in the habits, repertories and dispositions of bodies. 'Capacities to affect and be affected' are not, then, pre-discursive, in the sense of existing outside of signifying forces They are mediated through processes of agencement that involve but exceed the discursive. In the rest of this section I identify and distinguish this approach from two other partially connected but distinct claims as to why affect is non-representational: first, affect is rendered equivalent to a level of bodily life before representation and, second, affect is treated as a synonym for life's excessive generativity beyond representation.

Recent work on affect that has begun to think with the bio and neurosciences treats affect as a non-representational modality in a particular way: affect refers to neurological and biochemical changes that are non-conscious and non-intentional. By way of engagement with research in the affective neurosciences, work has attempted to show how social processes work below the threshold of consciousness, drawing together the biological, neurological and cybernetic (Wilson 1998).[8]

eyebrow, mouth down, tears, rhythmic sobbing), fear-terror (eyes frozen open, pale, cold, facial trembling, with hair erect), shame-humiliation (eyes down, head down), contempt-disgust (sneer, upper lip up), anger-rage (frown, clenched jaw, red face). It is important to remember the uncertainty that characterises how Tomkins (1995: 73) makes distinctions. He stresses that 'in sharp contrast with the primary drives, there is today no consensus on what the primary affects are, how many there are, what they should be called, what are the conditions under which they are activated and reduced, and what is their biological and psychological function'. The openness of his system is exemplified in the discovery of a ninth affect between the publication of the second and third volumes of *Affect, Imaginary, Consciousness* ('dissmell') (Sedgwick and Frank 1995).

8 This work resonates with efforts in corporeal feminism to think the materiality of the body outside of either social constructionist or performative accounts of bodily substance or psychoanalytic accounts of identity formation (Colls 2013; Lim 2007). Most notably,

Leaving aside epistemological issues of how neuro and biosciences are enrolled into claims about the nature of physiological shifts (Papoulias and Callard 2010), what this tendency in work on affect does well is break with any assumption of a self-contained individual. Neurological and biochemical processes are mediated through complex interrelations between a body and other forces. For example, Connolly (2002: 203–4 n7) highlights how bodily changes emerge 'out of the intersection between external events and the system of re-entrant combinations in a system like the body/brain network'. Addressing a similar problematic, Brennan (2004: 3) is concerned with the 'transmission of affect' as a 'process that is social in origin but biological and physical in effect'.

Consequently, attention is paid to the minutiae of neurological and biochemical changes and how those material changes express a subject's participation in processes; the blush that overcomes a face, a punctual shock, a vague sense of aliveness. Although supposedly inseparable from acts of conscious deliberation, judgment and decision, these changes occur at a different level than thought and reflection. They are, as Massumi (2002a: 28) puts it, 'irreducibly bodily and autonomic'. The wager is that there is a level of bodily life that people may not be wholly aware of, but that nevertheless has effects and can be and is being shaped or conditioned (in a way that leads us back to the recent concern with the manipulation of 'affect itself' discussed in section 2.2, see Barnett (2008)).[9] One of the ways in which bodily autonomic responses are mediated – as Connolly's (2002) emphasis on techniques of the self shows – is through practices of thinking refracted through media apparatuses that attach, or attempt to, an affective charge to particular forms of selfhood.

In this work, affect is non-representational in a specific sense; the neurological and biochemical processes of the body happen before the realm of what Leys calls (2011: 437) 'intentions, meanings, reasons, and beliefs'. Representation here means acts of conscious deliberation and discretion by a constituted subject through which that subject comes to interpret a world that they are involved in. The pre- or non-conscious level follows a different logic to consciousness. It is a domain of complexity, where proto-thoughts and proto-feelings emerge and pass away and where neural materialities have emergent qualities that escape subjects. For Connolly (2002: 27), for example, 'thinking and judgment are already well under way before they enter the picture as conscious processes' because 'affective charges help to move thinking and judgment in some directions rather than others'.

Grosz (2005: 194) argues for an engagement with (non)organic life through a 'politics of imperceptibility, leaving its traces and effects everywhere but never being able to be identified with a person, group or organization'.

9 Of course, this question has also been central to ideology critique. For this reason, critics of the 'affective turn' have, rightly I think, identified a kind of correspondence with ideology critiques in that both concern themselves with how subjects are manipulated before conscious awareness, in particular through the operations of the mass media (see Barnett 2008).

In related terms, Thrift (2007: 7) highlights the preconscious, emphasising how 'This roiling mass of nerve volleys prepare the body for action in such a way that intentions or decisions are made before the conscious self is even aware of them'. Hence the iconic role of the 'half second delay' in work on the materiality of perception (Connolly 2002; Massumi 2002a; Thrift 2004). For it is in the gap between the onset of action and the organisation of action that not yet determined or to be determined tendencies emerge and may take form. This gap is not, however, somehow pre-social or unmediated or natural. Instead, material perception is provisionally and partially organised through what Connolly (2011: 49) terms a 'history of inter-involvement' involving 'embodiment, movement, body image, touch, sight, smell, language, affect, and colour'.[10] It may also be organised through apparatuses that act on the gap between 'infra-perception' and 'perception'.[11]

There are clear commonalities between this emphasis on the preconscious and what Woodward and Lea (2010: 165) term the 'visceral immediacy of bodily encounters' that I began this chapter with; not least a shared ethos that avoids reducing all of affective life to subjective emotions, is attentive to emergent order from multiple processes, and aims to sense the complexity of perception. All theories of affect are also at pains to stress the historicity of affective life and its specific modes of incipient organisation and reorganisation (whether through the geo-historicity of encounters, the unconscious filtering of proto-thoughts or through techniques that intervene in the gap between the preconscious and the conscious). Nevertheless, beginning from encounters rather than a claim about the preconscious dimensions of affect does lead to some differences in orientation and attunement. For me, the starting point is that affects are transpersonal rather than pre-personal. The distinction is important. The former term – transpersonal – attunes us to how bodily capacities are mediated through forces that exceed the person. The key question is how does a body's 'force of existing' emerge through encounters, or more specifically the press and presence of the multiplicity that make up an encounter. Emphasis is placed on how many different bodies mediate affective life

10 The use of Benjamin Libert's experiment by prominent affect theorists has been subject to extensive critique (Leys 2011; Papoulias and Callard 2010). The charge is that scientific results are selectively used that conform to pre-existing ontological-political positions. Those studies are then used as evidence for the ontological-political commitments, a move that draws upon and repeats the emerging institutional and cultural power of the life sciences. Whilst I signal my disagreement with work that equates affect with a pre-discursive bodily substance outside of consciousness, I think critics, in particular Leys (2011), do not pay enough attention to the specific encounter with brain science and how exactly brain science is being used. Nor does Leys (2011) pay enough, or indeed any, attention to how meaning and signification are being thought about differently in recent affect theory.

11 Barnett (2008) shows the link between the ontologisation of a gap between action and reflection that subjects are unaware of and the political diagnosis that corporate actors and the political right now operate at the level of affect, understood as neurological or biochemical bodily reactions.

in relation. The focus is not on affect as some kind of material bodily substrate and its independence from signification and meaning (contra Leys 2011: 433). Rather than specifying the preconscious foundation of experience, the task is to attend to the mediation of what a body can do in specific social-spatial formations. On this use, the term affect is a 'weak' one, in Latour's (1998) sense. It orientates inquiry. Affect attunes us to what a body can do in emergent orderings and how 'forces of existing' vary, rather than supposedly ineffable physiological shifts in the body outside of consciousness.

Let us return to Fanon's example of the affective states of racialised subjects to illustrate how such an account might make sense of how encounters are mediated. First, the affects of race do not simply 'belong' to the encounter. They also 'belong' to the processes of racialisation that mediate the encounter and to past encounters between coloniser and colonised. Hage (2010) shows how the encounter is overdetermined by a racial schema that works, first, to particularise ('Look, a Negro!') and then, second, to ascribe negative characteristics to the now racialised particular ('Mama, see the Negro! I'm frightened'). The encounter works through a distinct process of racialisation that Hage (2010) terms 'racial mis-interpellation.[12] Whilst Fanon describes the immediate intensity of the encounter in a way that recalls accounts of psychic trauma, at the same time the encounter is inseparable from other times and spaces. He writes:

> I was responsible at the same time for my body, for my race, for my ancestors. I subjected myself to an objective examination, I discovered my blackness, my ethnic characteristics; and I was battered down by tom-toms, cannibalism, intellectual deficiency, fetishism, racial defects, slave-ships, and above all else, above all: "Sho' good eatin'" (Fanon 1986: 112).

Second, affect is inseparable from action: a particular force of existing. What Fanon (and the child) can do shifts, sometimes dramatically, sometimes imperceptibly, through the encounter. For example, Fanon writes of a shattering experience:

> My body was given back to me sprawled out, distorted, recolored, clad in mourning in that white winter day (113).

Third, the encounter is populated with ideas and meaning. For Hage, central to Fanon's encounter on a train is what he calls 'a hopeful belief in universality' (Hage 2010: 126) that disintegrates as Fanon is hailed as a particular subject: 'All I wanted was to be a man among other men. I wanted to come lithe and young

12 Hage (2010) distinguishes racial mis-interpellation from two other forms of racism: non-interpellation (where the racialised subject is rendered invisible) and negative interpellation (where the racialised subject is ascribed negative characteristics). (For recent work on the affects of racism and racialisation see Swanton 2010; Lim 2010; Saldanha 2007; Wilson 2011.)

into a world that was ours and to help to build it together' (112–13). Later Fanon (114–15) writes of a movement between a longing-for recognition and violent rejection: 'I shouted a greeting to the world and the world slashed away my joy. I was told to stay within bounds, to go back to where I belonged'.

Fourth, and perhaps in contrast to Deleuze's 'simple' example of a harmonious encounter between two people who greet one another, Fanon's encounter on the train is fractured. It involves misrecognition. It involves violence. The encounter does not proceed by way of alignment and coordination:

> the Negro is mean, the Negro is ugly; look, a nigger, it's cold, the nigger is shivering, the nigger is shivering because he is cold, the little boy is trembling because he is afraid of the nigger, the nigger is shivering with cold, that cold that goes through your bones, the handsome little boy is trembling because he thinks that the nigger is quivering with rage (114).

The participants in an encounter do not necessarily form a unity or community.

We can read Fanon alongside Deleuze as directing attention 'horizontally' to the multiplicity of different bodies that come together in encounters. The difference with the literature that draws on neuro or biosciences is, then, a difference of emphasis: crudely, between attuning to the transpersonal or the preconscious. This requires that we look a little closer at the claim that a body's 'charge of affect' can be said to be transpersonal. Whilst encounters may generate something different as elements come together, Fanon teaches us that encounters do not float free from spatially/temporally extended relations, nor are they immediate. Elements from elsewhere or elsewhen will be active participants in how an encounter happens, for example processes that particularise and then objectify the racialised individual are part of Fanon's example of the encounter on a train. Acknowledging that encounters are spatially and temporally extended is not, I would stress, to invoke some meta-structure through which affective life is organised in advance, or lazily to invoke power as an all-purpose context that pre-exists and determines life (chapters 2 and 3 having argued for the need to differentiate between modalities of power). Emphasising the entanglement of encounters in other relations, processes and events does, however, preclude beginning from the immediacy of the punctual encounter. It does not preclude, however, acknowledging the immediacy of affects. Indeed, Fanon centres on the facticity, the there-ness, of (post)colonial affect and the intensification of racism: 'I was responsible at the same time for my body, for my race, for my ancestors' (1986: 112).

For example, we can understand what was offered in chapters 2 and 3 as an attempt to understand one set of processes through which encounters are mediated. Whether by accident or design, attempts to shape, mould, manipulate and otherwise act on affective life become part of encounters and shape relations. Consider Foucault's (1977) account of disciplinary techniques. Discipline aims to produce a particular relation between obedience and docility: the body is made 'more obedient as it becomes more useful, and conversely' (138). By ordering cellular, organic, genetic and combinatory traits, disciplinary apparatuses aim for a particular augmentation and diminution of a body's 'force of existing'. What is augmented is the individualised body's ability to progress towards a perpetual end (what Foucault refers to as the 'organization of geneses'). What diminishes is the individuated body's capacity to resist the organisation of productive forces. In the case of discipline the core techniques of hierarchical observation, normalising judgment and the exam work to organise a multiplicity into cellular individuals. Productive capacities are extracted from life and the body is made useful. What disciplinary techniques also aim to do is ensure that bodies enter into encounters with accumulated patterns of action-reaction, as well as simultaneously producing those encounters. Avoiding any hypodermic model of the manipulation of affect, we should stress that apparatuses do not simply determine encounters. Instead, we can think of attempts to act on and through multiplicities as just one form of mediation that become part of encounters alongside many others. We might note, for example, how disciplinary techniques – such as the production of an 'organic individuality' through the temporal elaboration of acts – work against the ever present threat of idleness and how they are accompanied by the production of boredoms (along with a range of other new bodily capacities).

Drawing the concept of apparatus into relation with the concept of encounter is one way to avoid a kind of presentism in our analysis of affective life. Encounters extend temporally and spatially beyond their immediate taking place. Fanon's encounter on a train with a child is mediated by and mediates two processes of (post)colonial racialisation: negative interpellation and mis-interpellation. But it is also mediated by the particular bodies that momentarily come together on a train, including a young boy sat with his mother perhaps returning home, perhaps going elsewhere: 'Mama, see the Negro! I'm frightened! ...' (1986: 111–12). It is not hard to see, then, how the concept of encounter has become such an important part of the attention to the dynamics of affective life. Beginning from encounters attunes analysis to how affects are constantly mediated in and through relations; orientating inquiry to the heterogeneous groupings that shape what a body is and can do, paying attention to the material complexity of life, and being attentive to the surprising juxtapositions through which something new might be produced. Whilst, at the same time, analysis remains centred on how 'forces of existing' are inseparable from the ongoing, never completed, ordering of encounters – including how signifying apparatuses striate and become inseperable from encounters: ' ... Frightened! Frightened! Now they were beginning to be afraid of me' (Fanon 1986: 84).

Beginning from encounters means rejecting a second way in which affect may be said to be non-representational: affect as a synonym for life's exuberant generativity. This extends a now familiar point about the limits of any approach that focuses only on apparatuses – that life exceeds the intentions of those who aim to act on it – to make the claim that it is through affects that other ways of living and being are disclosed.[13] For affect is 'unqualified intensity' or a 'non-conscious, never-to-be conscious autonomic remainder' (Massumi 2002a) that always exceeds its expression or qualification, or so it is claimed by Massumi in particular.[14] What Massumi offers is a claim that affect is non-representational in that it is a point of view on 'the edge of the virtual, where it leaks into the actual' (43).[15] Invoking affect as a 'point of view' on the virtual leads to a contentious assumption that the expression and qualification of affect in feelings or emotions can only be a secondary frictional process of 'capture' and 'closure' that always misses affect. Thus, outside of an analysis of any particular encounter or specific affective complex Massumi claims that:

> affect is autonomous to the degree to which it escapes confinement in the particular body whose vitality, or potential for interaction, it is. Formed, qualified, situated perceptions and cognitions fulfilling functions of actual connection or blockage are the capture and closure of affect (2002a: 35).

Here the overspill, and remainder, of affect is assumed to constitute an opening to the virtual difference that is present in any situation or context. Elsewhere Massumi (2002a: 217) stresses the transversal qualities of affect that mean it 'is situational: eventfully ingressive to context. Serially so: affect is trans-situational'. Folded into what becomes actual is always a qualitative remainder of newness that exists outside of any specific determination: 'the remainder of ingressive potential too ongoing to be exhausted by any particular expression of it' (248).[16] Through the escape *and* ingression of affect, life is given 'the unseen possibility of other strange possibilities' (Rajchman 1988 cited in O'Sullivan 2001: 133).

---

13 This resonates with a broader critique of governmentality-based perspectives for their implicit behaviouralism: the assumption of some work on governing affective life that an apparatus simply (re)produces a particular form of subjectivity, or produces subjects who produce their own subjectivity in conformity with the needs or intentions of the apparatus.

14 The basis to Massumi's account of the excess of affect is the concept of the 'virtual', although the account of the virtual is aligned to a theorisation of the affective body in movement. The concept of the virtual stands in a family resemblance to a set of post-structural concepts that attune to an excess that grounds a radical affirmation of difference: including 'will to power, *différance*, Lack, (non-)being or ?-being, the body without organs, the "unsayable something", the differend, the feminine' (Widder 2000: 117, italics in original).

15 Hence the invention of new techniques for witnessing and fostering affective life – including those drawn from the performing arts and experimentations with modes of writing – and the reworking of standard techniques, most noticeably interviews and diaries (see Dewsbury 2009; McCormack 2003).

16 Although Massumi (2002a: 8) does stress how 'social and cultural determinations feed back into the process from which they arose. Indeterminacy and determinacy, change and freeze-framing go together'.

This account of an impersonal, excessive, life is doubtless appealing, perhaps even consoling in the midst of recent diagnoses of the manipulation of 'affect itself' (discussed in Chapter 2). It reminds us that life is never simply an object of or for power, and that something about life exceeds the many ways in which life is known, rendered actionable and intervened in. As such, it connects work on affect to flourishing attempts to develop an alternative approach to life that breaks with biopolitical frames and logics and invites an experimentation with alternative presentational styles.[17] Affective life is not just a sterile secondary product or outcome of some form of transcendent organisation or immanent arrangement. Nevertheless, the risk is that an ontological claim about 'affect itself' tells us little about how specific capacities to affect and be affected emerge and change as encounters are mediated. Claims about affect in general come to be grounded ontologically through the figure of the virtual, rather than, in affective neuroscience, through claims about a biological substratum accessible by the sciences (at times the two come to be equated with one another). Empirical work on affective life then ends up repeating the claim it begins from: that life is unattributable, exceeds any and all actual determinations, and always–already harbours the possibility that the here and now might be otherwise and perhaps better

Starting from an expanded understanding of encounters as mediating and mediated, means suspending ontological claims about the excessive nature of 'affect itself', just as it involves bypassing claims about the nature of an ineffable biological substance. To stay with how affects happen in encounters, I find inspiration in Eve Kosofsky Sedgwick's (2003) simple axiom cited in Chapter 1, aligned to Latour's (1988) principle of irreduction cited in this chapter's epigraph. Although Sedgwick centres attachment as one particular type of relation, we can read her as affirming the wonderful diversity of (non)human things that affects are mediated in and through but are irreducible to:

> Affects can be, and are, attached to things, people, ideas, sensations, relations, activities, ambitions, institutions, and any number of other things, including other affects (2003: 19).

Paraphrasing, we could say that affects *may* attach in surprising ways. The conditional is important, as it avoids grounding a guarantee that social life will be different and otherwise in a claim about the biological or virtual basis of affect itself. For what I have tried to stress so far in the book is that there are all manner of ways in which a body's 'charge of affect', its 'force of existing' is produced and reproduced. Capacities to affect and be affected may be formed through a geo-historicity of encounters, or the way in which space provides a setting and

17 Here I am thinking of attempts to cultivate what Esposito (2008) calls an 'affirmative biopolitics' that find in life a generativity and positivity that disrupt any attempt to draw a line within life between a valued life to be protected and a devalued or threatening life to be destroyed.

support for encounters. Apparatuses involve attempts to shape what a body can do in a given situation. Processes such as racialisation can fold encounters into other spaces and times. We shall see in the following chapters how structures of feeling and affective atmospheres act as conditions within which encounters happen and people feel. Affects are inseparable from these and other processes of mediation, with the result that, as Fanon's slashed joy reminds us, affects may constrain and restrict as well as enable, open up and disrupt.

4.4 Hope and the Contingency of Affective Life

I have distinguished my approach from two versions of affect that would both treat affect as the non-representational object per se (cf. Pile 2010). First, affect does not refer to a level of ineffable autonomic bodily reactions that are non-representational in the sense that they are somehow inaccessible to subjects. Second, affect is not equivalent to the generative immediacy of an always overflowing life – an equivalence that would mean representation always fails in relation to affect. Instead, my emphasis has been on understanding affect non-representationally, where affect refers to bodily capacities indistinguishable in practice from emotions. This means tracing how affects are formed in the midst of encounters and are mediated in ways that are not reducible to 'representational-referential' systems. On this understanding, there is no such thing as 'affect itself' and affect is no more the non-representational object per se than a table, money or a global pandemic. We can, in short, say nothing about this mysterious substance affect 'in itself' or 'as such'.

But Fanon's destroyed hope – 'I shouted a greeting to the world and the world slashed away my joy' (1986: 114) – returns us to the openness of some encounters. For his 'shouted greeting' implies that (post)colonial affective life was not wholly determined. How, then, can we hold together a sense of the contingency of affective life without grounding the openness of affect in a claim about life's generativity or biology? I address this question indirectly through two hopeful moments: Steve listening to music in the context of despair, and Emma listening to music in the context of grief. Both moments are from research with households into how music became part of everyday life.[18] Steve and Emma's experiences and words have stayed with me. It was to do justice to their experiences and others like them that I first became interested in theories of affect. Underpinning my attunement to the geographies of hope is the infralanguage specific to affect elaborated through this chapter; hope as a capacity that emerges in encounters; hopefulness as a

18 The moments are drawn from my previous research on music and everyday life with UK households that used a series of methods to attune to how music functioned in relation to the domestic geographies of affect, feeling and emotion, including observant participation, repeat individual and group interviews, periods of listening with and situated diary work (Anderson 2004a; 2005).

constellation of specific bodily background feelings emergent from the expression of affect; and actual hopes that become part of encounters through processes of qualification and are distinguished by possessing a determinate object. Considered as sensitising devices, rather than claims about the nature of 'affect itself', my wager is that these different terms allow us to understand the richness and complexity of the type of experience we give the name hope to. In short, hoping is something more than an future-orientated act in which it is only the content of what is hoped for that is socially and culturally mediated (see Nunn 1996; Waterworth 2003). Rather, and as set out above, encounters involve multiple processes of mediation and it is in relation to these processes of mediation that 'forces of existing' vary and encounters happen.

Hope has long had a connection to questions of how new possibilities emerge. The presence of hope has long been thought to herald a more-to-come, an excessive overspilling of life, that hints or gestures towards new forms of life. For Marcel (1965: 86), for example, being and becoming hopeful involves a 'radical refusal to reckon possibilities' (see also Marcel 1967; Pieper 1994). Hoping may accompany, enable or shelter possibilities for new forms of life. In the two examples of hopeful moments, however, it appeared that hope emerged from encounters that involved some form of loss: a body that became hopeful typically held 'the condition of defeat precariously within itself' (Bloch 1998: 341). Moments of hope were emergent from various relations that appeared to diminish.

> The first case refers to Steve who is 29, lives alone, and had just been made redundant from a job of two years with a manufacturing firm that has relocated out of the area in which he still lives. We talk about his hopelessness in the context of his job loss.
> Steve: " ... just been a bit bored and lonely ... everything's closed around here ... I'm not doing anything at the moment"
> Ben: " ... Yeah"
> Steve: " ... Sometimes I don't have the energy to do much else ... just sit here ... sit here too much and watch the world go past and see how shit the neighbourhood has become ... too many boarded up houses ... or go out and it's the same"
> We then talk, and listen together, to music as the interview progresses. He plays an album by Radiohead in the context of a discussion about hope:
> Steve: " ... I listen to this album ... in the morning and err ... if I'm feeling ... you know, like if I'm a bit low ... this'll cheer me up ... get me a bit more hopeful again. I don't know, it's quite melancholy ... but it's really beautiful ... and it stirs up emotion ... and I find it quite difficult to cry as well, and you need to at time to time ... so ... I ... this helps, so if I'm a bit low ... or I'm just lying here ... this is the song to put on ... umm"
> Ben: "Have you any other songs like this?"

> Steve: "'Mayonnaise' by the Smashing Pumpkins which is a very, very beautiful song … even though it's called mayonnaise (laughter) … a sad song called mayonnaise … what about you, what have you got?"
> Ben: "I don't know … radiohead creep was always one …"
> Steve: "Radiohead is just much more, MUCH more … just self loathing, which is … I always find it … a solace … to know someone else is feeling the same … which is great. I listened to this album yesterday morning when I'd been feeling down … it helps because I know I'm no way as bad as Thom Yorke but he kind of feels the same … it let me get on with it a bit …".

Until the final words hope is a trace here, a fleeting presence that moves in and out of the conversation between Steve and me, occasionally animating the talk. Witnessed instead are a set of diminishments within the present expressed through a series of bodily affections ('*bored, lonely, no energy*'). Hinted at, as Steve talks, are one specific set of diminishments that express the feelings of hopelessness and despair that may be set in motion by the event of unemployment. Steve discusses, poignantly, the frustration that marks simple practices of looking out of his window and walking through his neighbourhood after a period of economic change; encounters that are mediated by movements of transnational capital and waves of deindustrialisation. This is the first point of divergence within different hopeful encounters: the varieties of diminishment that condition an imperative to hope but also, in their differences, indicate that the capacity to hope is not evenly distributed (see Hage 2003). Steve's words resonate with, but are not equivalent to, the unequal distributions of hopeful encounters that mark the affective geographies of suffering more broadly. To give a very different example, Lasch (1991: 81) describes the persistence of hope for emancipation during the period of slavery in the American south despite the grave suffering and systematic violence. In the context of the specific affectivities of injustice bound up with slavery, hope existed as a counter 'belief in justice: a conviction that the wicked will suffer, that wrongs will be made right' (compare with Kumar (2000) on hope and the Zapatistas' struggle in Mexico, or Povinelli's (2011: 103) discussion of hopeful sites – 'sites where a potential alternative social project exists in the actual world' – in situations of extreme neglect within the shadows of late liberalism).

There is therefore a point of danger, or hazard, folded into becoming hopeful that indicates that a good way of being has 'still-not become': in the sense that 'the conditions that make it possible to hope are strictly the same as those that make it possible to despair' (Marcel 1965: 101). In Steve's case, it is from within the context of specific diminishing encounters that becoming hopeful emerges. Hence, perhaps, the sense that hoping abandons the existent and how some types of hope can also feed back to continue relations that diminish even as we are attached to them (see Potamianou (1997) and Berlant (2011) on 'cruel optimism' in the present). The transitions that Steve describes at the end of our conversation provide an alternative case in which becoming hopeful momentarily folds into a better way of being. It involves the expression of a movement of affect that counters a set of feelings ('*cheers me up, gets me a bit more hopeful again*') through an encounter with music. Momentarily, music smoothes over despair and induces the affective presence of something better, in this case by presenting a shared experience ('*to know someone else is feeling the same*'). Attuning to the lyrics and tone of Radiohead offers a hope that disrupts, momentarily, the circulation of despair through the ingression of something different and perhaps better into Steve's everyday life ('*it's quite melancholy ... but it's really beautiful ... and it stirs up emotion*'). This is less an alternative social project (cf. Povinelli 2011), and more a minor variation in a 'force of existing', one that comes and goes, brought into being for a while, before fading as other affects happen

Perhaps induced by the encounter between Steve and the music, and then him and his environment, is a bodily disposition of hopefulness (rather than necessarily the determinate content of a particular future that is hoped for). Hopefulness happens as a repeated set of background feelings that form a different relation to the future. For a while, a better future is affectively present. Hage (2003) invents the term 'conatic hope' to describe a specific constellation of background feelings that is irreducible to the ideational content of what is hoped for. Hage argues that certain types of hopefulness are akin to a sort of force of existing, linked to Spinoza's (1910) description of a finite mode's endeavour to persist in being (the conatus). For Hage, conatus is not the property, or essence, of a thing but a characteristic way of connecting and disconnecting that enables finite bodies to repeat across processes of emergence. Gatens and Lloyd (1999) draw on the idea of conatus to think through how a body is formed through transpersonal lines of force that augment or diminish and come together in particular encounters:

> Our bodies are not just passively moved by external forces. They have their own momentum – their own characteristic force for existing. But this is not something that individuals exert of their own power alone. For an individual to preserve itself in existence, as we have seen, is precisely for it to act and be acted upon in a multitude of ways. The more complex an individual body, the more ways in which it can affect and be affected by other things (Gatens and Lloyd 1999: 27).

When thought through this transpersonal account of the bodies capacities, types of hopefulness can be characterised as a continuation of good relations that enable an individual 'to act and be acted upon in a multitude of ways' (Gatens and Lloyd 1999: 27). Hopefulness therefore exemplifies a disposition that provides a dynamic imperative to action in that, in some cases, it enables bodies to go on. As a change in what a body can do, its 'force of existing', hoping opens the space-time that it emerges from to a renewed feeling of possibility: a translation into the body of the affects that move between people during encounters to make a space of hope. In Steve's case it facilitates the possibility of his being able to *'get on with it a bit'*. It is therefore this force of existing, hopefulness as a capacity to affect and be affected, that is the second point of differentiation between moments of hope.

We can witness the animating effect of different dispositions of hopefulness in cases such as illness (Parse 1999) or activism (Turkel 2004) when hope disrupts, if only momentarily, the lived experiences of suffering, harm or damage. Situations that had appeared to offer no way out are opened up, even if only momentarily, even if only in minor ways, and even if possibilities are then curtailed, lost, damaged or destroyed.

There are multiple intersecting dispositions of hopefulness that emerge from encounters with something, here music, which opens up the here and now. This has been hinted at by the examples I have invoked of additional cases in which hopes become part of some very different geographies (political struggle, illness, and so forth). Hope may therefore have a contradictory place in relation to everyday life. The diminishment that provides a kind of affective imperative to welcome and be open to a good future is itself called forth from the encounters that make a life. The absence, or desperation, that can sometimes be part of hope is not merely a possession of the individual but is a question of how the emergence, movement, expression and qualification of despair emerges from and conditions encounters. But the subsequent production of a disposition of hopefulness begins from a discontinuity that enables other tendencies to become part of how encounters happen. Becoming hopeful is not simply guaranteed by the overflowing qualities of immanent life nor marked by a simple act of ideational transcendence. Instead, in moments of hope new relations happen through which something different and perhaps better might be momentarily disclosed and made affectively present. A moment of contingency, a moment of openness, becomes part of encounters. But the moment is imbricated with the transpersonal forces – in the next case of love and loss – that mediate the encounter.

> Emma is 33 and married with two children. In the period during which we met and talked Emma had traced her biological father. Two days before the second interview she learnt he had died a year previously. We are talking about the effects of this event when she describes taking her son to the home of the parents who adopted her:
> Emma: "In that depression there's a sort of flatness and a sort of … lack of animation … and a lack of any sort of sharp feelings … and, and … so … and music is something, it always … it has a sharpness for me, yeah (yeah) and there was just no place for it … this week … in that dullness (yeah) … and every now and again it just breaks through … like … so anyway … Monday about eight thirty taking Noel to my Mum and Dad's and he was badgering me … he'd been badgering me on Sunday for cheerful music, and umm … I felt that I could bear it and might actually take me mind off it … so I put the usual tape on … you know Gina G, and Hot Stuff by Donna Summer … I totally disengaged from it, I put it on for Noel, and it was really brilliant to see him dancing along but it made me feel … quite sad … cos I couldn't feel anything with it … I felt like there was this huge glass wall, or glass … glass ears … or … that I couldn't engage with him at all … it just felt … dead … I couldn't engage with him at all … so

> anyway got back home … flicking through the channels and I got some soul music, and I don't know … what it was … something I didn't know … proper old soul … and it cheered me up, it was brilliant … I felt better. I felt instantly better because of that … as I had been sort of weeping on my way to dropping off Noel when … like … Hot Stuff had been on … but this … it reminds me of friends and dancing and home and my other family … so … you know I'm not going to feel like this always".

In Emma's case, and Steve's, a renewed feeling of other tendencies and latencies emerges from a disruption in the pattern of broader affective flows. Emma describes a similar set of transitions in feeling and emotion to Steve. The hopefulness she describes opens up around a point of diminishment that emerges from relations with both her child and biological father to make the car a site enacted by the circulation of grief. The grief and flatness of depression that is part of the event of her biological father's death feeds into a momentary interruption of her relation with her son. She finds it hard to relate to him in the context of his day-to-day moods. Some music induces and escalates this disconnection so it frames her relation with the world by enacting a set of distinct feelings (*'it made me feel ... quite sad ... cos I couldn't feel anything with it ... I had been sort of weeping on my way to dropping off Noel'*). Later, after she has returned home, soul music slowly comes to induce and amplify a disposition of hopefulness through a disruption of the transmission of grief. Emma remembers a time and space that is still forthcoming by reliving the intensities of a past set of events. Perhaps the materialities of music induce a not-yet source of hope by placing her home into contact with its affective past in a way that enables a different future momentarily to be felt (*'it reminds me of friends and dancing and home and my other family'*). Perhaps, her encounter with the music interrupts other relations. Perhaps a felt past becomes a source of hope, for a while.

The third point of divergence that creates different space-times of hope is therefore a moment of discontinuity in which a threshold is crossed through the creation of an intensified connection with life (the 'glimmer' or 'spark' of hope). From a context of potential diminishment, where that diminishment is still present and felt, something happens to enable bodies to go on with a renewed openness. Benjamin (1969) provides us with an intensified sense of this moment of transition in the allegory of the electrifying 'spark of hope' that holds the capacity to draw past and future together in the explosive disruption of now time. Hoping is not only dependent on the exercise of an individual faculty 'but is carried under the intrusion of an outside' (Deleuze 1988b: 87). There is a kind of felt beginning that disrupts existing relations. To hope is therefore to disclose not-yet-become 'seeds of change, connections in the making that might not be activated or obvious at the moment' (Massumi 2002c: 221). The affectivities of the opening up of hope, a movement that may surprise thought, are ignored or considered to be illusory in the argument that hope indicates the absence of joyful passions (Nietzsche 1986). It is the discontinuity of the event of hope that has long been heralded as the mystery of hope by secular writers who value an ethic of hope. Lingis (2002), for example, has talked of it in relation to an act of forgetting the past. Hope, to quote Lingis (2002: 23), 'arises from a break with the past. There is a kind of cut and the past is let go of'. Ernst Bloch (1986: 174), slightly differently, thinks of hope as disclosive of 'dawnings on the front of the process' through which the not-yet is sensed and the here and now is opened up, even if only for a moment.

This does not mean that 'affect itself' is somehow always in excess. Even if we retain the assumption that affects are mediated through encounters, and that the expression and qualification of affect does not exhaust the totality of affective expression, there is also a need to remain open about the exact relation between affective life and that which is 'in excess' of determinate forms and structures. Otherwise the risk is that the escape of affect over existing determination is invoked in catch-all terms, given a positive value and becomes nothing but the latest iteration of a search to find and found hope in the living of everyday life. In contrast to Steve and Emma's examples of hope, we could think of how encounters are organised in ways that annul, foreclose or suspend possibilities as well as foster or affirm them. For example, Probyn (2000a: 139–41) describes how the transmission of shame in queer pride movements produces a 'back-and-forth movement of distancing' between bodies that translates into 'a heightened awareness of what one's body is and does'. The body that is shamed is marked by an awareness that one has trespassed proximity and so 'loses any pristine sense of its boundaries: it is bespattered and besmirched by its own actions' (140). One very specific movement of affect here emerges from disruptive relations of antagonism and contestation that distribute bodies into hierarchies of interest and obligation. In this movement an affect of shame gives bodies different capacities to affect and be affected, but it is both that which is marginalised as outside the measure of the proper and that which is redundant in the context of the functional. The emergence and movement of shame, in this example, calls us to think multiple types of excess that move beyond the surplus-remainder couplet that is in the background to some accounts of the excessive nature of affect, in particular Massumi's (2002a).[19] If we supplement this work with a Marxist literature on the affectivities of different types of labour then we can see that not only does the movement of affect take place through different types of excess, but the expression and qualification of excess varies as affect folds into types of relations organised through distinctive encounters. In the performance of different types of bodily labour, such as certain types of emotional labour in the family or certain

19 Massumi's (2002a) stresses that the 'more' of affect constitutes a *line* that always 'escapes' as it 'goes on'. The terms 'surplus' and 'overspill' are also used in addition to 'escape', to disclose the presence of the movement out of context of a continued 'more'. Writing on the qualification of the excess of affect, Massumi (2002a: 36) asserts that 'when the continuity of affective escape is put into words, it tends to take on positive connotations. For it is nothing less than the perception of one's own vitality, one's sense of aliveness, of changeability (often signified as "freedom")'.

types of manual labour, a surplus does not automatically open up a virtual 'vague sense of potential' (Massumi 2002c: 214). Instead, the creation of a surplus of feeling is qualified in exploitative and/or oppressive relations that differentiate bodies in unequal relations (Fraad 2000). This should remind us that the excess of affect cannot be invoked in general terms. It must be understood as emerging in concert with the ongoing mediation of encounters. In the cases of Steve and Emma, hopefulness emerges as a property of encounters with music in the context of music being the kind of ordinary commodified thing that is now present in a wide range of everyday situations. Present in both of their encounters was the pain of some form of ongoing loss. Steve and Emma come to the encounter with music with existing 'forces of existing', ones that have been formed through the pattern of past encounters and come to mediate, contextualise, delimit and organise the 'population or swarm of potential ways of affecting or being affected that follows along as we move through life' (Massumi 2002c: 214).

The disposition of hopefulness, or more precisely how new connections are established that disrupt a diminishing organisation of space-time, opens the present to transformation by disclosing a topologically complex space-time of the not-yet. Perhaps, following Taussig (2002), hope can be thought of as akin to a kind of sense. Emma, for example, names a future good that is desired and discloses, importantly, the source of help by which one can attain it. The outside is named as a set of tangible connections to old friends and family that, when induced through soul music, disrupts the affective trajectories that dominate her life and enable her momentarily to hope that her grief will be overcome in the future. For Steve there are other people out there, including Thom Yorke and perhaps me, who sense the despair he does and enable him to name not being depressed as a hope that changes, again momentarily, how he encounters his home and neighbourhood. The result might be thought of as sites of experience animated by flashes of what Bloch termed (1986: 221) 'anticipatory illumination' that disclose that 'the world itself, just as it is in a mess, is also in a state of unfinishedness and in experimental process out of that mess'. The disclosure of that which is not-yet is a point of differentiation between practices of hope, in that the identity of the outside varies, but it is also that which enables hopefulness to be spoken of as a specific type of bodily capacity linked to the openness of not-yet futures and to a 'spark' that may open up the here and now.

The effect of the event of hope, the calling-forth of an outside that is different from the present, may be a more or less durable change in ongoing experience. This process of change in the affective background is attuned to when people describe the atmosphere of a space or time as hopeful or hopeless as felt through their own disposition yet existing independently of it (see Chapter 6). In the cases of Steve and Emma hopefulness as a feeling has also simultaneously crossed over a threshold to be felt personally as an emotion. Through a process of resemblance and limitation the full range of available potentialities become determinate possibilities as a hope is thought and named (Deleuze 1991). As processes of expression and qualification mediate encounters, hope like other affects troubles a straightforward distinction between the conscious and non-conscious, between bodily intensity and subjective emotion, or between the impersonal and personal. Hope and hopefulness blur into one another, as people name a feeling, become attentive to how they may feel in a particular situation, or offer reasons for feeling a particular way. It is partly through these representational acts that hope and hopefulness are mediated by specific signifying apparatuses that articulate and organise what can be hoped for. We could note, for example, how capitalism is marked by hopes of upward social mobility, or how consumer culture functions as a machine for generating and circulating hope by working on definitions of what can be hoped for (Hage 2003). Relations that diminish and destroy are, however, still present and may be actualised in feelings of loss when the object of hope fades away or feelings of disappointment when the object of hope passes unrealised. The frequency of such changes in the capacity to affect and be affected indicate hope's precariousness: becoming and being hopeful involves an encounter with a future that is not-yet and may never be. The resulting lack of guarantee, the fragility of a hopeful body's openness to something better, is brought forth in the context of illness, for example, when trust is enacted against the evidence of corporeal suffering (Lingis 2002), or in protest cultures animated by a belief in the possibility of alternatives despite persistent inequalities (Parker 2002). Different as these examples obviously are, they all hint to the contingency of affective life. Hope and other affects are made and remade, come and go, in the midst of temporary, more or less fragile, encounters.

4.5 The Imbrication of Affect

> Nothing is by itself ordered or disordered, unique or multiple, homogeneous or heterogeneous, fluid or inert, human or inhuman, useful or useless. Never by itself, but always by others (Latour 1988: 161).

The difference that hoping may make – an opening up of the here and now to something not-yet – is an outcome of how hoping emerges from a set of relations between bodies and in the midst of specific types of encounters. It does not automatically follow from an ontological claim about the virtual or biology, nor from valorising hope as some form of transhistorical universal emotion. In the moments discussed here, hope is perhaps best understood as a relation of suspension that discloses the future as open whilst enabling a capacity to dwell differently, if only momentarily. Differences are, though, integral to the processes by which space-times become hopeful or hopeless. Hope is not just one thing. Diminishment, disjuncture, hopefulness, the creation of an outside, the enactment of trust and dependence, the creation of a specific determinate hope, the feeding back of hope into life and the movement of hope into other bodily capacities, are some of the ways in which hopes and hopefulness vary. Whilst an act of hope may open up encounters to something that is not-yet present, capacities to hope pre-exist encounters and are shaped through them. Actual hopes will also emerge and end as part of how encounters happen. We saw something like this in Fanon's encounter on a train. Fanon (1986: 86) writes of the violence of slashed joy, but also the presence of a hope for recognition, a hope that is disappointable and disappointed: 'I shouted a greeting to the world and the world slashed away my joy'.

Just as I argued in chapters 2 and 3 that techniques of power do not simply reduce affective life, here I want to stress the complementary point: that we should not assume the excessive nature of 'affect itself' over the ways in which life is organised. Instead, it is necessary to remain open about the relation between excess and organisation. Following Sedgwick (2003) and her simple but important insight about attachment, the starting point for analysis is best summarised as the *imbrication* of affect with multiple forms of mediation, involving but extending beyond the signifying apparatuses we glimpsed at work in Fanon's description of a racialised and racialising encounter. Now routinely used as a way of signalling a connection without specifying the form of the resulting pattern, the term imbrication moves attention away from the ontological priority of affect, its autonomy (cf. Massumi 2002a), towards how affects are always-already mediated (by materialities such as music, by encounters with a hoped-for future, by connections to a lost loved one, by a deindustrialising landscape, by the inquisitive, halting words of someone learning to interview ...).

Affirming the imbrication of affect allows us to separate the assertion that affect is transpersonal from an ontological or biological foundation. What sense does 'transpersonal' thereafter have? Obviously the implication is that affect is not exclusively a matter of an individual, seemingly natural, emotion. However, the prefix 'trans' is ambiguous because it does not always and everywhere mean 'anti' or 'not' or 'non'. Rather than instigate a dualism it has a more complex sense of being between, in the middle, in the midst of. On the one hand, affect is transpersonal in the sense that affects must be thought of as things in the world that are formed through encounters and relations that exceed any particular person or any particular thing. 'Capacities to affect and be affected' are always collective in the sense that they are forged in and through the encounters that make up the realm of everyday life. However, on the other hand, affects are personal in the sense that they are expressed *in* a specific person or specific thing and change in that process of expression and qualification. Affirming affect as transpersonal is an attempt, then, to avoid beginning analysis by partitioning affect into a subject or an object, or into valued or devalued categories of the personal or the impersonal.

If we start from encounters the 'affective turn' involves more than a claim about preconscious bodily dynamics at a level below conscious awareness. For me, the key point of the 'affective turn' is not to recognise a previously ignored, marginalised or downplayed dimension of neurological or biochemical life somehow on the threshold or below consciousness. Rather, the provocation and promise of the 'affective turn' is to take encounters as the generic social thing, and to describe how a body's 'force of existing' is mediated as encounters happen. Returning to the previous two chapters, and looking ahead to the next two, we can think through how processes of mediation are many and varied and riven by their own volatility as they are set up or emerge, coexist and change. Returning to the moments of hope that first sparked my interest in theories of affect – grief-music-everyday life and unemployment-manufacturing-music – we see that affects may be one element amongst others in an encounter, affects are composed in the context of durable patterns of encounters, and affects are transformed in relation to processes of mediation. This means that the dynamism of affective life is not primarily due to the sub-individual, self-organising, properties of a non-representational biological matter. Moving away from claims about the dynamism of 'affect itself' leads to a style of description that understands the mediation of affective life non-representationally, in the sense of happening through a range of immanent elements and processes not limited to signifying forces. This involves; tracing the relations between the heterogeneous elements that compose encounters; following how apparatuses condition, code and link otherwise diverse encounters; and mapping how, from within encounters, bodily capacities emerge and change. It follows that capacities to affect and be affected are never equally distributed, they are always a matter of patterns of encounters, material arrangements and, occasionally, surprising events that may usher in something new and different.

The emphasis on bodily capacities, or forces of existing, as emergent properties of encounters is only one version of affective life. It leaves something important out. Throughout the book there have been hints of how collective affects such as fear or paranoia may come together to form part of the conditions for life. We find hints of a similar almost environmental definition of affect throughout Deleuze's single authored writings and his joint writings with Guattari. In Deleuze's 24 January 1978 lecture on Spinoza, for example, affect is described as a 'life' or simply 'existing' and 'this is what it means to exist'. And, elsewhere, Deleuze writes of a 'fog of a million droplets' (Deleuze and Parnet 2003: 48) or the 'incorporeal vapour' (47) that accompanies events. This environmental definition of affect resonates with the ways in which we may feel the atmosphere of a room or be able to tune into the mood of the contemporary condition. In the next two chapters I introduce two concepts with which we can speculate about affect as environmental – structures of feeling and affective atmospheres. Through these concepts I discuss a third way in which affective life is organised that extends and deepens my emphasis on apparatuses and encounters: collective affects form part of the conditions for how affective life takes place and, as such, mediate *affectively*.

Chapter 5
Structures of Feeling

> We find ourselves in moods that have already been inhabited by others, that have already been shaped or put into circulation, and that are already there around us
> (Flatley 2008: 5).

5.1 Affective Conditions

In the previous chapter I argued that affects are collective: capacities to affect and be affected are always mediated in and through encounters. Any particular body's 'charge of affect' therefore carries traces of other bodies and both reflects, and contributes to, some form of complex, changing, relational field. Nonetheless, one risk of this account of bodily capacities is that analysis remains focused on individual bodies, even as claims about relationality are repeatedly made. The affective turn becomes, then, another way of shifting attention from the personal to the transpersonal whilst at the same time attending to the formation of subjectivity. As such, it resonates with other 'turns' over the last 20 years that centre human subjectivity in order to dissolve it into a wider field of forces, events and relations. Whilst these moves are appealing, they risk leaving something important out. Chiefly, such approaches have less to say about how collective affects come together to provide the grounds for encounters and apparatuses. By which I mean that affects are not only emergent from encounters or object-targets in apparatuses. Collective affects are also part of the complex *conditions* for other processes, events and relations. Affects become the environment within which people dwell, as well as being capacities or object-targets.

We can find hints to an environmental understanding of affects that blurs the line between individual and collective in work that has attempted to diagnose the contemporary condition in terms of a dominant emotion or set of linked emotions. Examples include popular and academic claims about the existence of a 'culture of fear' in liberal democracies or an 'age of anxiety' in late-capitalism (see Wilkinson 2001; Furedi 2006). More specifically, work has identified the saturation of contemporary liberal democratic societies by historically and culturally specific moods: 'spite' (Diken 2009), 'neurosis' (Isin 2004), 'shock-boredom' (Woodward 2009) or 'cruel optimism' (Berlant 2011), to name but four. These attempts to diagnose something like a dominant structuring of emotional and affective relations resonate with popular attempts to articulate the 'national mood' that characterises an era. Consider how the British historian Dominic Sandbrook (2011) characterises early 1970s Britain in terms of a pervasive 'sense of emergency' that crossed between the state and the textures of everyday life. By contrast, geographers or historians of the Cold War have identified the 'anxious urbanism' of the Cold War era (Farish 2011: 193), or the Cold War as a 'neurotic symptom' where 'fears of the communist threat mask and deepen anxiety over the development of American political life' (Dolan 1994: 69–70). Across these examples, we find hints that particular collective affects may condition life, have a longer duration than affects or emotions, and come to frame the world as a whole rather than be attached only to particular objects. Collective affects may also be vague and diffuse, not equivalent to a body's 'force of existing' emergent from encounters or an object-target of apparatuses.

Rather than dismiss such accounts as involving an inappropriate scaling up of a body's 'force of existing', in this and the next chapter I want to experiment with understanding collective affects as part of the real conditions for how life is lived. Perhaps more fragile and transient than other more obviously material conditions of life, collective affects nevertheless become part of the background of life and living. This is to explore the third of the three relations between affective life and modes of organisation that this book focuses on: collective affects as part of the conditions for the formation of encounters and for the life of specific apparatuses.

In this and the next chapter I focus on two ways in which collective affects condition: through *structures of feeling* and through *affective atmospheres*. As we shall see, these two terms orientate us to different aspects of life and raise slightly different challenges for a social and cultural analysis of affective life. In this chapter I explore the peculiar existence of 'structures of feeling' by showing how affective relations and capacities come to be mediated by collective moods. For example, in the midst of the 2007–2008/present financial crisis something like precarity may cohere in a televised image of a politician justifying uncertainty in the present to save an imagined future, or it might be sensed through a precariously employed individual's rush of anxiety about the possibility of losing their job. What characterises precarity and other structures of feeling is that they are forms of affective presence that disclose self, others and the world in particular ways. Structures of feeling mediate life by exerting 'palpable pressures' and setting 'effective limits on experience and on action' (Williams 1977: 132). The infralanguage I offer in this chapter is designed to sense how structures of feeling have real effects as distributed affective presences that disclose life.

5.2 Mediation and Collective Moods

In his wonderful book *Romantic Feelings: Paranoia, Trauma, and Melancholy, 1790–1840*, Thomas Pfau (2005) discusses the concept of mood in a way that provides a first approximation of how collective affects condition encounters. Based on a reading of Heidegger on how mood conditions our engagement with the world, for Pfau a mood 'subtends' (6) diverse domains of life, drawing them into relation and giving them a certain type of 'enigmatic coherence':

> When approached as a latent principle bestowing enigmatic coherence on all social and discursive practice at a given moment, "mood" opens up a new type of historical understanding: no longer referential, thematic, or accumulatively contextual. Rather, in its rhetorical and formal-aesthetic sedimentation, mood speaks – if only circumstantially – to the deep-structural situatedness of individuals within history as something never actually intelligible to them in fully coherent, timely, and definitive form (Pfau 2005: 7).

What intrigues me about mood as articulated by Pfau is that he points to how some kind of collective affect can be at once operative in giving an 'enigmatic coherence' across life whilst also subtending life by persisting in the background. Because it is both present and absent, mood can only speak 'circumstantially' whilst nevertheless hinting towards how people are anchored in the world and to others. Mood hints to what Pfau (2005: 9) nicely terms, glossing Heidegger on 'having-been-thrust', as a shared '[d]ispositional relation to the world logically prior to any sustained cognitive engagement or incidental affective experience'. In order to develop Pfau's emphasis on mood, I will work through two ways of thinking about how something like a collective mood becomes the presupposition for and medium of life – an 'age of x or y emotion' or a 'culture of x or y emotion' – in order to draw out a common problematic that they share and which has been in the background to the book up to now; the problem of how life is mediated affectively. That is, how do collective affects come to be organised in relation to 'extra-affective' forces and how do they then come to condition through something like a shared 'dispositional relation' that situates individuals? How are individuals situated affectively in ways that may be incoherent, vague and changeable?

5.2.1 'Ages Of'

In popular, media and academic accounts, epochs are often defined in terms of a dominant public feeling: an emotion that when named expresses something about what it feels like or felt like to live in that particular period of time. For example, liberal democratic societies in the shadow of the so-called War on Terror are characterised by an 'age of anxiety' or by an 'age of fear' (see Pain (2009) for an overview of this literature and a critique). Swyngedouw (2010), for example, diagnoses how 'fear' infuses the post-political frame that, for him, conditions politics today. Writing on the apocalyptism of environmentalism, he claims that: '"Fear" is indeed the crucial node through which much of the current environmental narrative is woven, and continues to feed the concern with "sustainability"' (Swyngedouw 2010: 217). The Cold War 1950s was a time of 'paranoia' in the USA and UK, to give another example. Perhaps the global financial crash reveals the contemporary age to be one of 'greed', and so on. Often implicit, the claim is that the same feeling infuses diverse public sites, organisations and groups. If we live in an 'age of anxiety', for example, this is because anxiety is manipulated by various actors, and because anxiety now conditions how a range of diverse topics can be encountered and engaged with (Wilkinson 2001). Whether anxiety, fear, overconfidence, despair or joy, the mood is taken to be public. Not only is it shared between people who may never have met, but it also provides something akin to the common ground for the taking place of public life, including modes of political speech, organisation and action.

The idea that discreet periods of time can be characterised in terms of a single, identifiable, nameable emotion leads to some important starting points for an account of affective conditions. It acts as a counterpoint to any tendency to see a body's 'force of existing' as an exclusively individual phenomenon. Individuals may be aligned with that collective mood, or out of sync (Ahmed 2004). More interestingly, by insinuating that affective capacities and relations are patterned, it also cuts the link between affective life and surprising attachments. Attention is instead paid to how public feelings are structured at the level of broader collectives (or a particular section therein), rather than presume that affect is necessarily synonymous with the happening of encounters and the surprise of social life. Affect structures such as 'age of anxiety' are, however, themselves contingent in the sense that they are historically specific occurrences with a beginning and an end. A period of time – hence the term 'age' – is the unit used by analysts to understand order within contingency. There is also an important sense that public feelings are themselves elements within the social that interact with others and as such have an efficacy (so are not only a medium that can be manipulated by purposeful actors).[1] Although rarely worked through analytically, in the background to most popular accounts of collective mood and some academic work is a claim that collective moods have real effects and are irreducible to other non-affective determinations.

For example, in writing of the contemporary condition as a 'time of fears', Bauman (2006: 133) stresses that:

> Once visited upon the world of humans, fear become self-propelling and self-intensifying; it acquires its own momentum and developmental logic and needs little attention and hardly any additional input to spread and grow – unstoppably.

Leaving aside for the moment that Bauman does not say much about the processes through which fear becomes 'self-propelling and self-intensifying', he provides us with an incipient sense that collective moods are part of the conditions of a society and may have an efficacy alongside more obviously material conditions, including

1 The literature varies on this. At times the attempt to specify a link between collective mood and extra-affective forces becomes a form of social determinism whereby the prevalence of a named emotion is explained by set of recognised and identified social factors. Caught in a logic of explanation that reduces the collective mood to a secondary phenomenon, the result is that the collective mood becomes equivalent to any other topic: something to be explained by something else. Alternatively, some work falls back on the vocabulary of manipulation criticised in 2.2 whereby negative moods are purposely manipulated by actors for political ends.

the apparatuses discussed earlier. Infusing multiple dimensions of life, the mood may come to acquire its own momentum, becoming self-reproducing. Fear may produce fear. Fear-induced protective actions can multiply possible threats and dangers, contributing to the generation of further objects of fear and resulting in further protective measures (Davis 1999).

Perhaps because of these useful starting points, the concept of an 'age' of x or y emotion raises a number of questions, all of which stem from taking a period of time (an 'age') as the analytic unit for thinking about how collective affects produce something like a 'dispositional relation' across multiple domains that situates individuals. First, how can we avoid homogenising a society into one emotion that is taken to be dominant? Rarely is an account given of how different emotions become dominant and coexist. It would be very easy, for example, to characterise liberal-democratic societies in terms of an age of rage, an age of boredom, an age of a vague feeling of being connected, an age of pleasure in the suffering of others, and so on. A convincing case could be made for any and all of the above collective moods saturating liberal-democratic societies. Second, how exactly does a dominant emotion function within society? Is an emotion dominant at the level of lived experience and/or at the level for a society's representation of itself, both or neither? Very often, the emphasis is on how a society talks about itself, and what is followed are shifts in the language that surrounds a topic which is then used as evidence for the predominance of a mood. Finally, even if we did start from the idea that there are different 'ages' understood through a model of temporal succession, a whole host of secondary theoretical and methodological problems questions suggest themselves. What are the boundaries of an 'age'? How does one 'age' become another?

Given these questions, the virtue of the idea of an 'age of' is primarily in what it orientates us to rather than its use as either a descriptive or explanatory concept for analysis: it allows us to remember that collective moods are organised, that they have an efficacy, that they may cut across diverse domains of life, and that they will be articulated with apparatuses (recall Foucault's (2008: 67) connection between 'fear of danger' and liberal apparatuses). It also hints that at certain moments it becomes easier to discern the operation of a shared disposition relation and that a collective mood can be sensed through how it mediates capacities to affect and be affected.

5.2.2 'Cultures Of'

Many of the initial orientations to collective affects that I take from the meso-level idea of distinct 'ages of' can also be found in the partially-connected idea that a society is made up of one or more distinct 'cultures of' particular emotions. Similarly pitched at a meso-level, here the unit is no longer an ill- or undefined period of time, but rather the more complex level of distinct cultures.[2] Straight

2 In one popular-academic version, the term culture has been used interchangeably with the idea of an 'age of' to homogenise how a society or nation interprets emotions and

away this formulation has the advantage of allowing for the possibility that there may be a plurality of cultures with varied relations with one another, and that these cultures may cross historical periods of time, span different domains of life and frame diverse areas of life and living. Focusing on distinct 'cultures of' also provides a different answer to the question of how to understand order given the contingency of how collectives feel: collective moods will change alongside changes in how meaning is patterned and bodies are distributed in hierarchies of value. On this understanding a 'culture of' x or y emotion is simultaneously: (a) that which mediates between an individual and the world thus conditioning how things show up, and (b) that which is organised through some form of system of signification.[3] Culture is here defined in 'structuralising/signifying' (Seigworth 2006: 108) terms as a medium that conditions what is felt and what can become the focused, intentional object of an emotion. The basis, as Grossberg (2010: 182–9) has carefully argued, is a specific 'Euro-modern' version of culture as a realm or domain in which 'we first represent the world and then take up residence in our representations' (185, citing Carey (1989)). Culture is that which mediates our immediate, unthought, experience of the world. It is through some form of structured signifying mechanism[4] that immediate lived experience is patterned, comes to be meaningful and frames the world for a subject.

For example, in relation to debates about a 'culture of fear' the concern is with the signifying practices whereby an object comes to induce fear in relation to a subject positioned as fearful. The fearful subject is an effect of a distinction between his/herself and that object of fear.[5] Signifying systems and practices pre-position

---

interprets emotionally. Implicitly, the proposition is that there exist a defined number of cross-cultural named and already known psychological emotions (fear, shame, love, and so on). These are made into universals. Historical specificity exists, but it relates to the content of the emotion (what is feared or what is shaming) and which emotion comes to dominate in a given society. The dominant emotion then comes to define how things are interpreted and thus plays a framing role. Suffering from many of the same problems as the idea of distinct 'ages of', the additional problem with this approach is that the form that collective moods can take is placed outside of history and geography (see Moisi 2009). There is no space for the emergence of new collective moods that stand outside what have become the normal taxonomies of experience.

3 This is one of the versions of culture that have coexisted in the 'new' cultural geography, alongside the emphasis on culture as a 'whole way of life' we find in someone like Peter Jackson (1989). As Seigworth (2006) expertly traces, its origins are in the paradigm of culture studies derived from what Stuart Hall (1980) termed the 'structuralists' in distinction from the 'culturalists' of Richard Hoggart, Raymond Williams and E.P. Thompson (not a term the latter three authors used).

4 There are of course significant differences in the nature of this mechanism, including ideology, specific uses of the terms discourse or representation, cognitive interpretation and narrative structures (Grossberg 2010: 187).

5 More recently in the post-9/11 context, see Hopkins (2007), Dwyer, Shah and Sanghera (2008) or Puar (2007) on fear and the racialisation of the figure of the 'terrorist'

particular individuals and classes of people so that their appearance or threat of appearance may induce fear. One consequence is that various harmful or damaging attempts are made by groups to counteract the now threatening object of fear. For an example of this process, consider Stuart Hall and collaborators' classic work (1978) on mugging and policing; still one of the best examples of an analysis of the relation between collective mood and representational mechanisms and schema. Through their study of the emergence of the problem of mugging they show how fear was attached by politicians, the state and the media to particular racialised bodies (young black men living in the UK inner cities in the late 1970s/early 80s). The existence of a moral panic around mugging was part of the precipitating conditions for the emergence of what Hall (1988: 27) called an 'authoritarian mood' identifiable across the spheres of law and order, education and welfare (a mood that, for Hall, was central to the 'authoritarian populism' of 1980s Thatcherism). Put in the terms of the previous chapter, we could say that 'representational-referential' mechanisms are one way in which particularised bodies arrive at encounters with an already specified 'capacity to affect' others and potentially 'to affect'.

There are some things this version of culture does very well. Chiefly, it attempts to connect up individual lived experience with what we could call extra-affective forces, by specifying distinct processes through which collective moods emerge, are cultivated and sustained. It also disavows us of any appeal to a body's 'force of existing' as a pure, authentic, domain to be counterpoised to representation. Capacities to affect and be affected are, on this account, mediated by a range of signifying mechanisms. As work on the relation between social differences and affective life show, these include representational categories and frameworks that order particular kinds of bodies in hierarchies of worth and value. Collective moods are not only felt differently according to an individual's capacities to affect and be affected, but are also made through gendering, racialising and other processes that operate through signifying apparatuses (Puar 2007; Ahmed 2004).

These insights come at a cost, however. That cost is a divide between signification and world, between a system of signification that conditions and a lived experience that is conditioned (Anderson and Harrison 2010). To overcome this divide theories that begin from a 'signifying/structuralising' version of culture have to identify mechanisms whereby the system of signification is (re)produced through individuals (and thus generate particular 'capacities to affect and be affected'). Let us look at two such mechanisms – 'feeling rules' and 'emotional discourses' – that offer an account of how a dispositional relation is formed. Both have been central to the burgeoning sociology of emotions and have been taken up in geography and elsewhere. Both begin from classic sociological questions about collective affects: How can that which feels most individual be shared by distant others and how does the ordering of feelings relate to the (re)production of social orders? In offering responses to these questions, the concepts of 'feeling rules' and 'emotional discourses' shift analysis from how collective moods are shared

(in particular in relation to Muslims).

to how an individual's feelings and emotions are formed through a 'signifying-structuralising' system of signification.

The concept of 'feeling rules' was introduced by Arlie Hochschild (1975; 1979; 1983) to understand how 'culture' 'directs action' at the level of feeling. In one of her first discussions of the term, she describes 'feeling rules' as the 'underside' to ideology – where ideology is understood as the 'framing rules' that ascribe meaning to situations (1979: 557). The concept is an answer to a classic sociological question of order:

> Why is the emotive experience of normal adults in daily life as orderly as it is? Why, generally speaking, do people feel gay at parties, sad at funerals, happy at weddings? This question leads us to examine conventions of feeling ... [which] become surprising only when we imagine, by contrast, what totally unpatterned, unpredictable emotive life might actually be like at parties, funerals, weddings, and in the family or work life of normal adults (Hochschild 1979: 552).

Hochschild's (551) answer is that 'emotion is governed by social rules ... ' and that those social rules have an 'imperial scope'. Drawing on Erving Goffman, she argues that 'feeling rules' '[g]overn how people try or try not to feel in ways "appropriate to the situation"' (552). They do so through a set of shared conventions that impinge on an individual's capacity to try to or try not to feel – that is their ability to undertake 'emotion work' to try and change the degree or quality of an emotion (562). 'Feeling rules' are, then, guidelines that direct the extent, direction and duration of feeling given the situation: 'These rules come to consciousness as moral injunctions – we "should" or "shouldn't" feel this, we "have a right" or "don't have the right" to feel that' (Hochschild 1983: 69). Like a script or guideline, 'feeling rules' can be slavishly followed or improvised, they can break down, we can be reminded of them, and we can wilfully or otherwise break them. A space of freedom is, therefore, opened up because 'feeling rules' stand in-between an individual's 'emotion work' and a social situation (one that resonates with the discussion of the excess of affect in the previous chapter). The key question for Hochschild (56) is, then, tracing the: '[v]arious ways in which all of us identify a feeling rule and the ways in which we discover that we are out of phase with it'.

The concept of 'emotional discourse' (Lupton 1998) is a little different, owing more to a mid-1990s translation of Foucault's term 'discourse' into a system of signification than the interplay of psychoanalysis and the interactionist sociology of Garfunkel that is in the background to Hochschild's original work. But Lupton does place the same emphasis on the interpretation of feelings and the (re)production of social order that Hochschild does. Focusing on 'discourses of emotion', as defined by Lupton (1998: 25), directs attention to '[h]ow we interpret bodily sensations and represent them to ourselves and others as "emotions"'. Immediately we should note a distinction between two orders; what Lupton term 'bodily sensations' and a representational economy. Instead of conventions which take the form of imperatives structuring how individuals try and try not to feel with regard to specific situations, discourses are societal-wide representations that work to structure what can be thought about the emotions and the expression of emotions. In terms that relate back to my discussion of versions in Chapter 3, there are, for example:

> [S]everal dominant discourses related to the ontology of emotion and how it is conceptualised in terms of selfhood. These included: emotion as a means of self-expression; emotions as signals or messages of one's thoughts or feelings; emotions as a means of self-authenticity; emotions as a personal resource; emotions as the seat of humanity; emotion as an essential part of life; emotion as primitive; emotion as the opposite to rationality/reason (Lupton 1998: 69).

To understand how discourses operate one must, therefore, identify the range that exist and their interrelations, as well as trace how they are cited in specific situations, whilst also ensuring that they are 'contextualised' and 'located' within the social structures of a particular society. Whilst seemingly more fluid, and multidimensional, 'discourse' is used in a very similar way to how Hochschild uses the term 'ideology' – to name a system of signification that pre-exists lived experience and provides a frame for the ascription of meaning to emotions (a use that is quite different from Foucault's understanding of discursive apparatuses).

There are obviously some dangers in running these two accounts together. They involve different versions of the subject who feels, for example. Hochschild's account makes her/him into an 'emotion manager' able to work on his or her inner emotions (and the outer display of emotions). Lupton makes the subject who feels into a secondary effect of a discourse that can be (mis)cited. Notwithstanding these differences, we can understand both as solutions to the dilemma of understanding how collective affects are patterned if we begin from a separation between a signifying system and individual subjects that are pre-positioned in that system. In both the presumption is that what is shared between individuals is not feelings per se. What is shared before feelings is a pre-positioning of subjects in a system of signification. People share a predisposition to feel and to interpret those feelings in accordance with rules or discourses. This starting point allows the analyst to show that the unmalleable aspects of feeling can be social or cultural, by invoking a source of determination that exceeds any specific situation and yet structures what can be felt. As both Hochschild (1983) and Lupton (1998) argue there has been a tendency to equate the biological with the fixed or invariant and the social with change and variation. Biology somehow acts as the raw material that is to be managed or worked on. Broadly social constructionist accounts of collective mood ask us to reverse this focus by assuming that the social is an ordered phenomenon and that, following on, feelings express the ways in which the social is ordered. Both 'rules' and 'discourses' are therefore mid-level concepts that aim to solve a set of dilemmas that accounts of a simple 'age of' x or y feeling pass over: what is the relation between an individual and external events? How does the individual act in relation to a set of conventions? And how is structure reproduced at the site of encounters, episodes and situations? In answer to these questions 'rules' and 'discourses' allow the analyst to specify how a body's capacities are the site for the reproduction of an already-existing pattern, without assuming a correspondence between individual feeling, collective affect and social structures and conditions.

What is most useful about the idea of a 'culture of' x or y emotion is, then, the focus on a specific process of mediation that complements and extends the emphasis on the collectivity of mood that we find in the idea of an 'age of' this or that emotion. By reproducing a Euro-modern version of culture, the emphasis is on signification as one specific mechanism through which an individual body's 'charge of affect' comes to be organised at the same time as collective mood takes place. It is also by revealing the operations of systems of signification – ideologies or discourses – that what is felt to be personal can be revealed as a symptom of particular collective relations and modalities of power.[6] What this work does so well, therefore, is to remind us that lived experience is never autonomous. Whether by following rules, or citing discourses, collective affects are connected to extra-affective forces. In doing so not only does this work break with any residual biological determinism[7] in the treatment of affect and emotion, but it also forces us to consider how dispositional relations may be part of encounters, become part of life as it is lived, and have an efficacy. Nevertheless, we have to avoid equating mediation in general with one particular form of mediation: the formation of affects, feelings and emotions in conformity with a system of signification that is 'mirrored' (Massumi 2002b: xiv). The problem is that introducing an intermediate level of mediation (the rule or discourse) assumes the existence of two pre-constituted entities: an ostensive structure through which meaning is given to experience and individuals. Collective moods are reduced to systems of signification. What exists are systems of signification (a 'culture of') that get reproduced through 'rules' or 'discourses' and merely mirrored in a collective mood.

Surprisingly, in the accounts of an 'age of' x or y emotion we have a more complicated, if mostly implicit, sense that a collective mood can itself be a form of mediation that conditions a body's 'capacity to affect and be affected'. Here we

6 There is a set of questions that follow from this emphasis on mediation as signification. For the main part, these go unaddressed in the literature. How should we understand the genesis of the 'thoughts and ideas' that stand in-between the individual and society? Through what processes or practices does a 'feeling rule' or 'discourse' emerge, change and end? Likewise how is the system of mediation lived and encountered – whether it is a 'rule' being followed or a 'discourse' being cited?

7 Despite the emphasis on the biological by certain affect theorists (as discussed in Chapter 4), I nevertheless think that 'biological determinism' remains a problem, notwithstanding Sedgwick's (2003) point that social theory cannot be equated with distance from biology. I define 'biological determinism' as the use of a claim about the invariant nature of human species-being to explain a particular action, process, encounter or event.

might think of the *reality* of a collective mood in the sense that although ephemeral, and indefinite, a collective mood can come to affect other entities. This poses some challenges though: how do we think the reality of something ephemeral like a collective mood? How do we think the reality of collective moods alongside a concern with signifying apparatuses and other forms of mediation? And how might we learn to consider collective moods as part of social life, mediated through signifying apparatuses amongst other ways, but also themselves conditioning life? In answer to these questions, this chapter returns to an interruption to the version of culture that accounts of mediation as signification have been based on. With regard to affective conditions specifically, this alternative version finds inspiration in the cultural materialism of Raymond Williams and his attempt to attend to those aspects of affective life that exist 'at the very edge of semantic availability' (Williams 1977: 134) but nevertheless form a set of 'pressures' and 'limits' through which life is lived. Here we find a more complicated sense of the organisation of collective moods that are themselves elements within life rather than reducible to signifying systems. To offer such an account is, however, simultaneously to outline a different account of how 'capacities to affect and be affected' are imbricated in other processes. One that starts not from two separate worlds that are then drawn together – lived experience and the ascription of meaning to that lived experience – but from learning to attune to what Williams (1977: 128) described as 'forming and formative processes' that involve but are not reducible to signifying processes and systems. This takes us back to the previous chapter and supplements the emphasis on affect as a bodily capacity emergent from assembled encounters between bodies. Structures of feeling exert pressure and set limits on how encounters can be felt. Following on, and slightly differently, structures of feeling are not only mediated through signification and representational devices. Collective moods are both mediated and themselves an active form and force of mediation.

5.3 Structures of Feeling

In *The Long Revolution* Williams (1961) characterises a middle-class structure of feeling as expressed in the popular fiction of the 1840s. His discussion of 'instability and debt' provides a good example of what the concept of structure of feeling might enable us to focus on. Williams begins by characterising the 'middle class social character' of the time: a set of ideals and values around piety, thrift and sobriety. Juxtaposed to this 'social character' is a structure of feeling that is not reducible to those ideals and values:

> The confident assertion of the social character, that success followed effort, and that wealth was a mark of respect, had to contend, if only unconsciously, with a practical world in which things were not so simple. … What comes through with great force is a pervasive atmosphere of instability and debt (Williams 1961: 65).

'Instability and debt' – and the anxiety that they will befall people regardless of their character or effort – are akin to conditions of existence that is 'pervasive'. Those conditions may be expressed differently in personal feelings and emotions, and people may be wholly or partially aligned with them or out of sync, but they exert a force on life as it is lived and delimit what can be experienced. On this understanding, a structure of feeling is a collective mood that exists in complex relation to other ways in which life is organised and patterned, without being reducible to those other ways (hence the difference between a structure of feeling and what Williams describes as the 'social character' of an epoch or ideologies). A structure of feeling is also particularising: 'instability and debt' is inseparable from a certain middle-class form of life, or what Williams (1961: 65) terms a 'practical world'.

The 'deliberately contradictory' (Williams 1977) concept of structure of feeling was first introduced by Williams in *A Preface to Film* (Williams and Orrom 1954), used in *Culture and Society* (Williams 1958), developed in *The Long Revolution* (Williams 1961), and receives its most explicit elaborations in *Marxism and Literature* (Williams 1977) and the career retrospective interviews that followed that work, *Politics and Letters* (Williams 1981). Unsurprisingly, given that it cuts across Williams's 'left-Leavisite' phase and his latter engagements with Marxist literary theory, the concept changes (Simpson 1992). Particularly important for my account of collective affects is how the concept oscillates between being a concept of mediation – standing in between individuals and structures – and a concept for a totality – albeit an expressive totality linked to a particular version of 'culture' made up of 'specific indissoluble real processes' (Williams 1977: 82). In this way the concept can take forward what is useful in the ideas of an 'age of' or a 'culture of': that collective moods have a real effectivity and as such mediate how 'capacities to affect and be affected' emerge, the focus on the workings of mediation, and the attention to collective processes of organisation. And 'structures of feeling' are collective because they are shared between people, dispersed and distributed across sites and come to condition emotions and feeling.[8]

Williams (1977) places the concept of structures of feeling in the context of the subject/object distinction that discussion of collective affects is too often caught within.[9] The problem is that what is changing is often made into a property of the

8 Despite the increased use and engagement with 'structures of feeling' in the context of the 'turn' to affect and emotion (see Boler 1999; Woodward 2009; Harding and Pribram 2004) the concept remains elusive and writers often fall back on very general definitions. For example, for Boler (1999: 28): 'Structures of feeling name the simultaneously cultural and discursive dimension of our experience but do not neglect those experiences that are embodied and felt'. Harding and Pribram (2004: 870) offer the following more careful definition: 'Structures of feeling as mediating concepts are specific deployments of emotion at specific historical junctures with particularized effects'.

9 Williams's emphasis on the 'lived' and 'felt' has been critiqued – sympathetically and carefully – by some on the basis of its supposed distance from traditional analytical

subject and that which is fixed gets made into the property of society. Two poles: subject and society. One living, the other dead. What is missed are 'forming and formative processes' (Williams 1977: 128) or what Williams terms the 'generative immediacy of social formations' (133):

> And then if the social is the fixed and explicit – the known relationships, institutions, formations, positions – all that is present and moving, all that escapes or seems to escape from the fixed and explicit and the known, is grasped and defined as the personal: this, here, now, alive, active, "subjective" (129).

We see the same problem in attempts to specify 'a culture of' x or y emotion: a distinction between a representational-referential system that provides something akin to the conditions of possibility for what can be felt and subjects who creatively or not relate to that a priori system. The problem is with assuming that the condition is fixed and acts by (pre)positioning subjects who then respond in idiosyncratic ways. As with various non-representational theories, Williams's aim is to think an open-ended social that is in process and ongoing (see Harrison 2000). That is a social composed of entities – such as structures of feeling – that are not fixed and explicit, yet still have palpable effects on how life is lived, felt and organised. This means that the interesting question becomes how does a structure of feeling have effects and become part of the social? How does a structure of feeling mediate?

In the context of the subjective/objective problem, and a protest against reducing the social to fixed forms, structures of feeling are defined by Williams (1977) as 'social experiences *in solution*'. The same phrase – 'solution' – is used in one of the few places in *The Long Revolution* (1961) where Williams offers something close to a description of the concept. The mistake is in taking each 'element' in a society as fixed and definite. This leads to a difficulty for social analysis that Williams unpacks through terms from chemistry:

> We learn each element as a precipitate, but in the living experience of the time every element was in solution, an inseparable part of a complex whole. The most difficult thing to get hold of, in studying any past period, is this *felt sense of the quality of life at a particular place and time*: a sense of the ways in which the particular activities combined into a way of thinking and living (1961: 63, emphasis added).

Whilst the problem is the same, and I will return to thinking the social 'in solution' below, the definition of a structure of feeling is a little more general than it will become later on. Here it is equivalent to the 'felt sense of the quality of life' and

---

scholarship and its Leavisite methodology (see Simpson (1992: 24) on its 'vitalist-empathic character' or Eagleton (1980) on its 'romantic populism'). Simpson (1992: 15), for example, makes the mistake of equating structures of feeling with a 'subjectivist emphasis' in the relation between the personal and general.

'a way of thinking and living' (Williams 1961: 63) in a defined period. This makes the concept particularising: it is the 'sense' or 'way' in a *particular* time and place. Practically, this means that Williams links structures of feeling to some form of collective organisation. The collective organisation – principally a generation or epoch – comes to stand in for geo-historical specificity.

For example, in *Culture and Society* Williams (1958: 119) identifies a 'general structure of feeling' expressed by George Eliot in *Felix Holt* that was 'the common property of her generation'. Likewise in *Modern Tragedy* (Williams 1966), three very broad structures of feeling are identified by Williams: 'mediaeval', 'bourgeois' and 'liberal'. Later in *Politics and Letters* (1981: 166), he characterises a 'middle class' structure of feeling of the time in terms of 'an anxious oscillation between sympathy for the oppressed and fear of their violence'. As we see from these examples, at times Williams links structures of feeling to distinct 'classes' (1977: 134) and at others to 'generations' (1958). Whilst both avoid over-generalisation at the level of society, there is a key and important difference between them. Linking structures of feeling to distinct generations (or periods of time) gives the impression that any given age can be said to have a single dominant emotion. It can be subject to the same critiques I made earlier of the idea of an 'age of' a named emotion: affective life is homogenised and reduced in a model that assumes linear succession. The link to class is more interesting because it ties structures of feeling to the composition of the social, provided that we do not presume beforehand that class is the only social formation that structures of feeling become part of the conditions for or are mediated by.

Reflecting on Williams's linking of structures of feeling to class and generation, Flatley (2008: 27) notes that unlike more general terms such as 'mood', 'structures of feeling' has the virtue of orientating us to specific formations. Flatley gives the following examples of some named 'structures of feeling' that link to specific groups, generations or classes. Unlike Williams, Flatley does not only link structures of feeling to class or periods but also to other collectives; residents, members of protest groups, races, and so on:

> For example, depression is a mood, not a structure of feeling; however, we might describe the particular depression of the Russian peasant in the steppe in the 1920s as a structure of feeling, or the depression of the residents of a decimated New Orleans after Katrina as a structure of feeling. Or, to return to an earlier example, we might talk about the structures of feeling created by the civil rights movement and the Black Panthers, structures of feeling that were mobilised within the *Stimmung* that allowed the 1967 rebellion against the police in Detroit to happen (Flatley 2008: 27).

This is useful, but as well as raising the question of how exactly structures of feeling are 'created' it also leaves a little opaque exactly what a structure of feeling is. How might we supplement the discussion of collective mood as disclosive by treating 'structures of feeling' as part of the ongoing affective composition of common worlds?

In the most explicit treatment of the term that Williams offers – in *Marxism and Literature* (1977) – we find the same concern with a way of life 'in solution' as we find in the earlier *The Long Revolution* (Williams 1961), but definitions of the term 'feeling' multiply, extending beyond its earlier use to name a generalised 'sense' of a generation. Examples include: 'practical consciousness', or what is 'actually being lived'; 'a kind of feeling and thinking'; and 'a particular quality of social experience and relationship' (Williams 1977: 130–31). These various approximations make the category of 'feeling' more or less equivalent to the general category of experience (or at least certain uses of). At one stage Williams (132) describes 'structures of experience' as 'the better and wider word', not using it because of the then subjective associations of the term 'experience' and the sense that an 'experience' was something finished that a subject had gone through in the past.

Returning to Pfau's (2005) discussion of mood in light of Williams's terminological ambiguity, we can understand a structure of feeling as giving an 'enigmatic coherence' across different domains of life in two respects. First, a structure of feeling *is* the affective quality that is common across otherwise disparate practices, events or processes. By which I mean that a structure of feeling is one way in which a dispersed collective is gathered and comes to have some form of coherence, if only a temporary one. Second, a structure of feeling gives a kind of unity to a multiplicity through that characteristic affective quality that cuts across, draws together, and holds together disparate practices, events or situations. The term 'feeling' in the concept does not, then, simply refer to a scaled-up version of a psychological emotion or a physiological feeling. Rather, the term feeling in structure of feeling names a particularising, shared, affective quality that acts as a type of disposition towards oneself, others and the world and emerges alongside some kind of collective.

Structures of feeling might be defined, then, as collective affective qualities that dispose bodies. Returning to and rereading Stuart Hall and colleagues' classic account of mugging (Hall et al. 1978), we could understand the 'panic' that surrounded mugging as an affective quality intimate with the birth of the 'law and order society' in the 1970s. One that condensed around the 'folk devilish' form of the 'black mugger' and involved a racist disposition that, in the terms of Chapter 4, organised affective attachments to young black men and mediated encounters.[10] What Hall et al. show is that the panic itself, rather than the act of

10 Hall et al. (1978: 18) used Cohen's influential idea of 'moral panic' to understand the shift of attention from the act treated in isolation ('mugging') to 'the relation between the deviant act and the reaction of the public and the control agencies to the act'. Hall et al. are interested in how themes of race, crime and youth 'condensed into the image of "mugging"' (viii) and focus on the relation between how the state polices the phenomenon of 'mugging' and the emergence of what they term an 'authoritarian consensus' (or what they term a '"soft" law and order society').

mugging alone, comes to have a causal role in conditioning and determining the response of the state to the act of robbery. Propelled by fears of societal decay and social change 'condensed' (Hall et al. 1978: viii) into the image of 'mugging' and centred on young black men, repressive state measures are demanded by the media to allay another collective affect linked to media apparatuses – a crisis of public confidence in law and order. The moral panic about mugging becomes one of the affective conditions for Thatcherism, in particular the mood of 'authoritarian populism' Hall (1988) diagnosed.

In *Marxism and Literature*, Williams (1977: 131–2) gestures towards how structures of feeling, as collective affective qualities that dispose bodies, come to condition life when he stresses that structures of feeling 'do not have to await definition, classification, or rationalization before they exert palpable pressures and set effective limits on experience and on action'.. On the one hand, to 'exert palpable pressures' is for a generalised, dispersed, affective quality to develop a particular form of presence across different domains of life. For one example of this process consider Berlant's (2004: 10) intriguing suggestion that compassion and coldness are not opposites but 'two sides' of a bargain that some middle-class white subjects of liberal modernity in the West have struck with structural inequality. Using Williams's terms we could say that combinations of coldness and compassion provide something akin to the frame within which structural inequality comes to be made present to those who are not on the receiving end of inequality. On the other hand, to 'set effective limits' is to become part of the dispositions that enable experience and action (the impulses and restraints). We might think of how compassion and coldness might preclude other affective relations to the persistence of structural inequality; outrage, shame, anger. What we can do here is think about how a collective affective quality – a particular structure of feeling – can delimit how other people, groups and things appear and can be related to and valued. Compassion and coldness make up a shared disposition that comes to condition how suffering can be and is related to. A humanitarian structure of feeling coexists with other affective conditions. For example, in *Inventing Human Rights*, Hunt (2007) shows how the emergence of a shared collective affect of 'imagined empathy' for ordinary suffering was part of the complex conditions for the creation of human rights as a social and political concept.

For another example of how structures of feeling exert 'palpable pressures and set effective limits on experience and on action', consider Mark Fisher's (2009) searching diagnosis of 'capitalist realism': 'The widespread *sense* that not only is capitalism the only viable political and economic system, but also that it is now impossible even to imagine a coherent alternative to it' (Fisher 2009: 2; emphasis added). Linking this sense to social groups caught up in neo-liberalising apparatuses of rule and government, Fisher's account is particularly useful because he describes this sense (feeling in the terms used by Williams) as an affective complex that is part of the contemporary condition; as much made up of an anxiety about the loss of the capacity of the future to produce surprises, as the increase in diagnoses of various psychiatric and affective disorders, or the new forms of boredom that accompany types of entertainment capital. Importantly for our purposes here, Fisher shows how this 'sense' exceeds any one location. As a shared feeling its traces can be discerned across politicians' proclamations that 'there is no alternative', apocalyptic films such as *Children of Men*, spaces of further education and call centres, amongst many other events, sites and processes. Whilst we might argue with the details, the key point here is that the sense of inevitability crosses diverse sites. Diffuse, it acts at once to condition how life is lived and to delimit thought and action. Alternatives to capitalism come to seem absurd; the unlimited accumulation of capital is cast as inevitable and inequality legitimised. A structure of feeling, then, conditions how something appears – in this case capitalism – by organising the way in which it comes to be felt as part of the dynamics of everyday life. Returning to Pfau's (2005) intriguing discussion of mood, we could say that a structure of feeling is a collective dispositional relation to the world.

What we perhaps do not get a sense of from the example of capitalist realism, but is central to Williams's formulation of the term, is the dynamism of a structure of feeling. 'Structure' in this context does not mean an extrinsic source of determination that conditions by standing apart from and pre-determining practical life. A structure of feeling is instead a 'solution': an expressive unity in which an affective quality both crosses between and is produced through the interaction of a multiplicity of elements:

> Yet this specific solution is never mere flux. It is a structured formation which, because it lies at the very edge of semantic availability, has many of the characteristics of a pre-formation, until specific articulations – new semantic figures – are discovered in material practice … It is thus a specific structure of particular linkages, particular emphases and suppressions and, in what are often its most recognizable forms, particular deep starting-points and conclusions (Williams 1977: 134).

Leaving aside that it is unclear quite what Williams means by 'deep' starting points and conclusions', structure is given particular characteristics. It is differentiated from either a pure, undifferentiated, flux or a fixed, definite, set of relations. For Williams, it is like a 'pre-formation' in that it lies on the edge of semantic availability and is still in process and therefore characterised by continual change as it is articulated. Again, Hall and colleagues' (1978) analysis of mugging can be reread to illustrate the point. They show how a concern about street crime intensified into a panic about mugging in the midst of a series of reconfigurations in British society, principally the use of force and law by the British government to suppress challenges. 'Panic in relation to mugging' as a structure of feeling coalesced in relation to a series of challenges in the mid-1970s to the ability of the then elite to govern through consent (specifically, open warfare in Northern Ireland, growth in student militancy and the power of the unions) and the fractured, contingent, relations between the institutions who responded to that 'crisis of hegemony' (government, media, police and judiciary). What we have a sense of is how a structure of feeling mediates a wider political crisis whilst, at the same time, becoming part of the crisis and how it was governed.

Later, Williams offers a more specific definition of 'structure':

> We are then defining these elements as a "structure": as a set, with specific internal relations, at once interlocking and in tension. Yet we are also defining a social experience which is still in process, often indeed not yet recognised as social but taken to be private, idiosyncratic, and even isolating, but which in analysis (but rarely otherwise) has its emergent, connecting and dominant characteristics, indeed its specific hierarchies (Williams 1977: 132).

Williams shares something with the structuralist Marxism that he was in dialogue with, but also at pains to differentiate himself from: a structure of feeling is dispersed across differences and shapes but exceeds actual encounters. He also lays stress on the internal relations between different elements. However, his description of a set that is 'at once interlocking and in tension' is slightly different from a notion of structure as a set of internal relations that determines actual occurrences. It suggests that a 'structure of feeling' is not an organic whole in which different elements are seamlessly integrated. The relations a structure of feeling is composed of may contradict one another, or may resonate with one another. As a social formation 'in solution' a structure of feeling is a 'complex whole' made up of 'forming and formative processes' rather than 'formed wholes' (Williams 1977: 128). This is perhaps closer to a forgotten notion of structure as a term of process. Tellingly, Williams (1976: 258) in his *Keywords* entry on structure defines it as the 'activity of building and the thing built as well as (in and through) the modes of construction'. Emphasising structure as a term of activity, reminds us to focus on the dynamism of the dispositional affective quality that is a structure of feeling.

This means that a structure of feeling is always-already *emergent*. Because a structure of feeling lies on the 'edge' of semantic availability, what is apprehended in analysis are the many and varied traces of a forming structure of feeling (which Williams finds expressed and recorded in art and literature in particular).[11] Williams tends to treat structures of feeling as emergent in one particular way. Whilst he is not consistent on this, for the most part 'structures of feeling' as a concept is distinguished from terms used to describe the realm of already articulated and existing beliefs and values, in particular 'ideology', 'conventions' and 'social character'. For example, in *Politics and Letters* he describes structure of feeling in terms of 'all that is not fully articulated, all that comes through as

---

11 For Williams, art and in particular literature anticipates what may then become institutionalised or formalised in ideas. There are links here to Ernst Bloch's (1986) focus on the 'anticipatory illuminations' that art and literature can provide. We should note that there is ambiguity in Williams's account of whether it is content and/or form through which a structure of feeling becomes apparent.

disturbance, tension, blockage, emotional trouble' (Williams 1981: 168). Whilst in *Marxism and Literature*, structure of feeling is identified with the pre-emergent: the 'active and pressing but not yet fully articulated, rather than the evident emergence which could be more confidently named' (Williams 1977: 126).[12] In one of the few examples Williams gives in *Marxism and Literature* he articulates the relation between 'ideology' and 'structures of feeling' through the example of early Victorian literature and 'exposure':

> Early Victorian ideology, for example, specified the exposure caused by poverty or by debt or by illegitimacy as social failure or deviation; the contemporary structure of feeling, meanwhile, in the new semantic figures of Dickens, of Emily Brontë, and others, specified exposure and isolation as a general condition, and poverty, debt, or illegitimacy as its connecting instances. An alternative ideology, relating such exposure to the nature of the social order, was only later generally formed: offering explanations but now at a reduced tension: the social explanation fully admitted, the intensity of experienced fear and shame now dispersed and generalized (Williams 1977: 134).

In this case, although not necessarily all cases, a structure of feeling follows a formed and dominant ideology and precedes a new ideological formation. It is an intensity that whilst irreducible to ideology, occurs in the midst of existing and reforming ideologies. Not only, then, is a structure of feeling emergent in the sense of being in process and revealed through its manifestations, but it is also emergent in the sense of being secondary to how affective investments are organised in relation to the realm of the 'articulated'.

We must distinguish between two uses of the term 'emergent': emergent as 'in process' and emergent as 'secondary to the realm of the already articulated'. For the most part Williams uses the concept in the latter sense, whilst offering a broader version of culture that emphasises the former sense of life in process. My working presumption is that 'structures of feeling' can be emergent in the sense of; (a) being continually in process and therefore in excess of attempts to specify them, without being emergent in the sense of (b) being disruptive to a dominant organisation of signification that is historically if not ontologically primary. Otherwise the risk is that we privilege the immediacy of the 'actively lived and felt', much as some work on affect privileges the incessant dynamism of non-conscious bodily matter. To cut the link between structures of feeling and the pre-emergent is to wager that collective affects may be mediated by the realm of 'articulated' thoughts and beliefs and, moreover, may be one of the ways in which the 'articulated' is reproduced. Here we can find inspiration in the work introduced in section 5.2. For what work on 'ages of' or 'cultures of' affects/emotions shows

12 On this use, structures of feeling would be a supplementary term to some form of system of signification (whether named in terms of ideology or discourse, notwithstanding their differences).

is that signifying apparatuses mediate by organising grids of affective investments and distributing bodies (Grossberg 2010: 201).

By juxtaposing two 'seemingly contradictory' terms, the concept of structures of feeling provides a wonderfully suggestive supplement to the accounts of the mediation of collective moods that I briefly discussed in section 5.2. Most importantly, it treats structures of feeling such as 'capitalist realism' or 'austerity' or 'fear of others' as real phenomena in the sense that they condition and become part of how social life is patterned and lived. With the concept, affective conditions come alive: diverse structures of feeling are in process, always being articulated with all manner of social actors, having real and palpable effects throughout life, and being one of the ways in which bodily capacities are organised and mediated. Structures of feeling effect in a twofold sense in that they both 'pressure' and 'limit' our relation to ourselves, others and the world. They exist as both dispositional attunements and affective qualities. There is also strangeness to the idea: structures of feeling are emergent and revealed through their traces. They can never be fully summed up and accounted for.

In attempting to hold together a set of seeming opposites – the structural and the aleatory, the stable and the ephemeral – several problems emerge for analysis. The next section draws out some of those problems, and develops the implications of not relating structures of feeling solely to the pre-emergent, via a short example of what we could perhaps name as an emergent structure of feeling specific to neo-liberalising processes in late capitalism: precarity and precariousness. Before we do, we should note that my account risks hypostasising a 'structure of feeling' as a distinct thing in itself. The following section will play with that risk by employing a practice of speculation as I move quickly between particular expressions of precarity and an account of the general formation and functioning of precariousness. Because it 'hovers on the edge of semantic availability', articulating a structure of feeling is always a matter of speculative description. It is a matter of offering a 'cultural hypothesis' of changes that might or might not be underway. I speculate in order to attune to the *irreducibility* of collective moods and their *efficacy*: how an affective condition simultaneously mediates life and is always-already mediated.

5.4 Precarity, Catastrophe and Other Structures of Feeling

In a reflection on his experience of precarious work in temporary agencies, Ivor Southwood writes of the costs of flexibility, the pain of precariousness as he moves between temporary jobs, always on the verge of unemployment or underemployment:

> Work is no longer a secure base, but rather a source of anxiety and indignity, both a matter of life and death and utterly meaningless, overwhelming and yet so insubstantial it could run through our fingers. It is normal to feel under threat

> and undervalued, to feel snivellingly grateful to have a job, any job. We must be sure not to take work for granted and yet be willing to be taken for granted ourselves. … We can barely live independently now. How will we be able to bring up children, or support them in similar circumstances? The future is no longer something to look forward to, but something to dread (Southwood 2011: 76).

Southwood describes the precarious existence lived by many in a world of temporary work characterised by a pervasive insecurity. Avoiding nostalgia for Fordism, he nevertheless draws on his own experience of various temporary jobs to explicate the affects of working and searching for work amid insecurity. Opportunism, fatigue, restless activity and forlorn hope come together as he participates in a structure of feeling that he and we could name as precarity. Moving between the particular and general, Southwood's hypothesis is that precarity is common to diverse forms of work built on transience and involving the perpetual but vague sense of being on the verge of personal crisis. Instead of the promised future offered by the linking of the Fordist idea of the 'career' and the work contract with the availability of credit, in precarious living the future is lost. What exists is the endless time of the present characterised by restless activity in a desperate bid to navigate the insecurity that follows from the breakdown of various Fordist guarantees; namely security in employment and the welfare state. For Southwood, the future as such or in itself comes to be present through a generalised dread that conditions his life of endless short-term work interrupted by equally endless job searching. Unlike fear which attaches to specific objects, Southwood expresses a general instability. He describes this condition in his reportage from the frontline of precarious work: 'My partner and I exist in such a state of constant instability or predictable unpredictability, always on alert, trying to outrun our own built-in obsolescence. Uncertainty has become the usual state of things' (76).

And yet, Southwood and his partner still have dreams. Ordinary dreams:

> My partner dreams of a job which is not a temporary fixed-term contract and which takes her less than two hours of commuting each way; to have transport costs which are less than a month's rent (current monthly season ticket: £383); to live near enough to work to have a social life and for her working relationships not to be undermined by inter-departmental competition. … In the manner of a couple imagining some gleaming Utopia, we wonder what it would be like to own our own home, to go on holiday together, to reclaim the hours filled by commuting and job applications (77).

If precariousness is the making present of a 'predictability-unpredictable' future, it can also involve the reproduction of dreams for something like a 'normal life'. Where the normal life aspired to, or wondered about, might be animated by a vague nostalgia for stability and involves a desire to live more comfortably in the world. And when that normal life is tightly bound up with economic and media apparatuses that circulate and reinforce a hope for a kind of ordinariness that is perpetually on the horizon (Berlant 2011).

Southwood's explication of the relation between uncertainty and exploitation resonates with other attempts to bear witness to precarity in work; blogs that tell of the stresses of endless job searches; conversations about estrangements from roles; accounts of being at the risk of redundancy; management texts on how to motivate agency staff; films that draw attention to the pain of migrant labour; discussions of how to move seamlessly from one project to another, to name but some (on precarity see Neilson and Rossiter 2006, 2008; Berlant 2011; Ridout and Schneider 2012). From these diverse sources, we can speculate that precarity exists as a structure of feeling not so much on the 'edge' of semantic availability as subject to an endless array of (re)descriptions that name something in common across forms of work. Rather than the disenchantment and dehumanisation which marked the experience of work for some in Fordism, precarity is the name given for a collective disposition of generalised insecurity whereby the future becomes 'predictably unpredictable' (Southwood 2011: 75). Precarity is at once intimate and extended across the spaces and times of a post-Fordist world of work: a collective condition felt individually, to use Southwood's phrasing. For, as Sennett (1998: 31) stresses, '[w]hat's peculiar about uncertainty today is that it exists without any looming historical disaster; instead it is woven into the everyday practices of a vigorous capitalism'. Rather than change being an exception that abruptly alters an otherwise routinised, stable, normal life, precarity as a mood that pressures and limits is related to economic worlds in which 'instability is meant to be normal', or so Sennett (1998: 31) argues. Precarity is not a temporary aberration. It is a generalised affective condition.

How do structures of feeling such as precarity relate to other 'formed' dimensions of the social in addition to class and generation? How can we think the relation between, say, a structure of feeling and a new way of patterning the social? Williams is interesting on this, precisely because he leaves the relation between change and structures of feeling open. An example is the following passage which emphasises the complexity of relations and how they exceed attempts easily or glibly to account for a structure of feeling through some form of all-purpose explanatory cause:

> For what we are defining is a particular quality of social experience and relationship, historically distinct from other particular qualities, which give the sense of a generation or period. The relations between this quality and other specifying historical marks of changing institutions, formations, and beliefs, and beyond these the changing social and economic relations between and within classes, are again an open question: that is to say, a set of specific historical questions (Williams 1977: 131).

Williams poses a vital question here: what are the relations between differentiated structures of feeling as particular 'qualities' and all the other elements (institutions, beliefs, economic relations, and so on) that go into making up the social as a set of 'specific indissoluble real processes' (Williams 1977: 82)? What is clear from this passage and Williams's other writings is the contingency of the relations between a shared 'quality' and how the social is patterned and organised. There is no homology between collective and structure of feeling.

Returning to precarity, we can identify a series of partial connections between precarity and the apparatuses that make up post-Fordism. Precarity is related to apparatuses based on; decentralised forms of labour process and work organisation; the emphasis on choice and product differentiation; the contracting out of services and functions; and the maximisation of choice around consumption. These are not abstract forces, somewhere detached from and conditioning life. Precarity is one name for how they may be actually lived and felt as an affective quality, albeit in ways reflected, enacted and expressed by a body's existing 'force of existing'. For example, for those outside of the Fordist guarantee the precariousness of existence is not news. As Mitropoulos (2010) rightly and forcibly argues, for migrant labourers, domestic workers and many others precarity has been the norm under capitalism. The guarantees offered by participation in Fordism were only offered to some and were dependent on the precariousness of the existence of others. Perhaps we can say precarity names a structure of feeling specific to individuals or groups whose existence was previously marked by the promise of future stability. For example, we might say that part of precarity as lived by some people is a nostalgia for the Fordist guarantee, as we find in the repetition of what Berlant (2011) brilliantly identifies as 'aspirational normativity': an affective reinvestment in 'the normative promises of capital' despite the past failures of that promise. Moving between absence and presence, we should stress that residual structures of feeling associated with other apparatuses also become part of the contemporary condition. Nostalgia for the Fordist guarantee, for example, infuses into and blends with work now. Just as apparatuses integrate a differential field made up of a 'multiplicity of extraordinary diverse processes' (Foucault 2007: 238), so structures of feeling will be composed through a heterogeneous mixture including other dormant affective qualities. Whilst the idea of an 'age of' an emotion homogenises collective affective life, the concept of structures of feeling orientates us to coexistence – diffuse, vague, affective qualities may blur with one another. Precarity may coexist with a ressentiment directed towards those who have stable employment, for example: the mood that we should all be in a 'race to the bottom' together.

Whilst serving to name a dispositional relation common to otherwise separate forms of work and employment, precarity is not a term in wide circulation. Unlike many of the names we give to collective moods (fear, panic, and so on) precarity is not part of the familiar Anglo-American lexicon of named emotions.[13]

13 Here we could distinguish precarity as a question of labour and work from precarity as a condition that 'bestows an enigmatic coherence' across multiple domains of life. Likewise, we should distinguish this account of precarity as an affective condition

A neologism coined by European social movement activists in the early twenty-first century, precarity originally named a range of insecure, temporary, forms of work.[14] Otherwise separate forms of labour were supposedly linked by a shared position within a post-Fordist labour market: casualisation, short-term or no contracts, freelance and undocumented employment (Waite 2009). Over the course of the mid-2000s precarity became the name for an experience of work to be mobilised against due to the presumption of a shared position in relation to capital. Images of work, testimonies by workers and artistic interventions became ways of translating a inchoate experience into something nameable and actionable that could be mobilised around and from which new social movements might be born.[15] The argument being that the experience of forms of work constituted an emerging 'precariat' amid reconfigurations of post-Fordist capitalism and neo-liberalising apparatuses of rule.[16] From a term describing the emergence of the 'flexiploitation' associated with insecure labour conditions, precarity became a rallying cry. Invested with the hope of a new revolutionary subject and new creative modes of social organisation based around a common experience of insecurity across differences,[17] mobilisation around precarity temporarily coalesced in a range of mid 2000s activist events associated predominantly, although not exclusively, with the Euro May Day movements in 2005.[18] Through forms of transnational activism, precarity moves from being an experience to be resisted (equivalent to

---

from Butler's (2004) use of the term to attune to the fragility of (human) corporeal life lived in and through relations to others.

14 More specifically, these were forms of work which broke from the feelings of public status and recognition that Fordist work entailed, albeit only ever for some (Muehlebach 2011).

15 We can also think here of journalist and artistic attempts to render visible the diverse cultures of precarious work outside of the vocabulary of precariousness (Berlant 2011). Here the aim has been to render visible forms of labour that exist in the interstices, the hidden spaces.

16 In this chapter I connect precarity and precariousness to post-Fordist capitalism and neo-liberalism, whilst acknowledging the sheer range of descriptive terms used for the contemporary economic-political transformations in labour that precarity has been linked to. I should be clear that this does not mean that precarity is somehow a new collective condition lived affectively. As Neilson and Rossiter (2008) and Mitropoulos (2010) argue it is Fordism that begins to look like the exception. Rather, it is to connect an affective quality and dispositional relation to, on the one hand, post-Fordism as a set of value-producing activities and, on the other hand, neo-liberalism as a set of modes of reasoning and techniques. The vocabulary of apparatuses and techniques offered in chapters 2 and 3 are in the background to this and the next chapter.

17 Central to the political use of the term precarity is an assumption that new modes of political organisation are necessary that are not reducible to the social state, Fordist economic structures or trade union solidarities (Neilson and Rossiter 2006: 1).

18 Other events included: Precarity Ping Pong (London, October 2004), the International Meeting of the Precariat (Berlin, January 2005) and Precair Forum (Amsterdam, February 2005).

terms such as alienation), to something more ambivalent: a site of pain and hope, loss and possibility. Even amid lives lived on the verge of crisis, the hope was that new forms of commonality might arise, emerging from the shared experience of uncertainty and the end of the guarantees (of rights and recognition) that followed from participation in the Fordist 'salaried society' (Castel 1991). For some, precarity might be the precondition for new forms of creative social organisation (Neilson and Rossiter 2006).

Precarity is not only, though, the name for a position in a post-Fordist labour market, or a political term to be mobilised against or to invest hopes for change in. What precarity names is an affective quality and dispositional relation: one that cuts across and links otherwise diverse forms of work amid post-Fordism and neo-liberalising apparatuses. It is in this sense that the term structure of feeling works well as it orientates us to a specific type of presence and how that presence comes to set limits on experience. Let us be more specific, then. My cultural hypothesis is that precarity names the affective presence of an unstable present and a disposition that discloses a 'predictably unpredictable' future without guarantees. For what defines the affective quality of precarity is not only that the present is saturated with a sort of restlessness, but also that the future is made uncertain and becomes difficult or impossible to predict. And what precarity names, then, is one mode of disclosing and relating to the future affectively. It is not only a labour market position. Rather than being present through the ominous heralds of catastrophe or the dream of progress, precarity makes present an unstable here and now vaguely menaced by an uncertain future.

In an interview published on 24 December 2004, the theorist and activist Paolo Virno (2004b) provides a way of developing this account of precarity as a quality and disposition that reflects, expresses and enacts the contingency of life. For Virno, precarity names a historically specific collection of anxieties and fears that resonate together to mediate relations to others and the world, one characterised by an uncertainty that oscillates between being a diffuse fact of life and occasionally intensifying around particular dangers.[19] Individual fears or anxieties are crystallisations of precarity, actualisations whereby a structure of feeling is intensified around particular futures to become part of capacities to act:

19 Although not the focus here, we should note that there is a series of connections between Williams's concept of structures of feeling and Virno's (2004a: 84) of an 'emotional situation'. In both there is an emphasis on something that is common across diverse contexts and how what is held in common can be affective. Virno explains the idea in the following terms in the context of a discussion of opportunism and cynicism in relation to post-Fordism: 'With the expression "emotional situation" I do not refer, let it be clear, to a cluster of psychological tendencies, but to ways of being and feeling so pervasive that they end up being common to the most diverse contexts of experience (work, leisure, feelings, politics, etc.)' (Virno 2004a: 84).

> It is a fear in which two separate things become merged: on the one hand, fear of concrete dangers, for example, losing one's job. On the other hand, a much more general fear, an anguish, which lacks a precise object, and this is the feeling of precarity itself. It is the relationship with the world as a whole as a source of danger. These two things normally were separated. Fear for a determinate reason was something socially governable while anguish over precarity, over finitude, was something that religions or philosophy tried to administer. Now, by contrast, with globalisation these two elements become one (Virno 2004b, no pagination).

Structured in the sense that it involves the holding together of specific types of fear and dread, precarity might exist as a vague, diffuse, uncertainty about what the future might be and bring or intensify in a focused sense that which specific futures threaten. Existence is made precarious in the sense that it is always on the *verge* of falling apart as specific dangers intensify from and become part of a diffuse affective condition.

On this understanding, the structure of feeling that I am terming precarity is not only formed in relation to particular types of employment, nor even the realm of work. Rather, precarity as a quality and disposition also subtends other domains of life that are connected to, but separate from, the concerns of work. Recall that Williams (1977) describes structures of feeling as 'social experiences in solution', precisely to emphasise the systematic and transient together. Speculating further, we could say that precarity in work resonates with the sudden and, for some, surprising recognition of contingency elsewhere in relation to post-Fordism and neo-liberalism: in the unwanted but frequent emergence of ordinary emergencies; in periodic crises; in a pervasive catastrophism that covers everything from climate change to terrorism; and in the return of apocalyptic framings of the future. Even though they share little else, in relation to these ways of disclosing the future life is framed as precarious and bound up with 'predictably-unpredictable' dangers brought forward from the future. For one example, consider how 'austerity', in many ways the replacement for the War on Terror, folds into a sense that economic life and the state are equally precarious, dependent on inhuman flows of finance, liable to fall apart and requiring continual monitoring and management.[20]

If we begin with a presumption that multiple structures of feeling coexist, how do 'structures of feeling' relate to one another and change – through sequencing, replacement, oscillation, contradiction or some other mode of relation? Williams's (1977) vocabulary of dominant, residual, emergent and pre-emergent poses this question, even if structures of feeling have been too closely identified with just one of those categories (the pre-emergent in the sense articulated above).

---

20 This passage draws on recent research on how life is governed through risk, including precaution, pre-emption and preparedness. What this work sketches out is how crisis, emergency, apocalypse and catastrophism exist simultaneously as structures of feeling, governmental techniques and ways of grasping and handling events (see Ophir 2007; Aradau and Van Munster 2011; Amoore 2011).

In addition to precarity, the predictably unpredictable future may also be made present and disclosed through catastrophic or apocalyptic structures of feeling. A catastrophic or apocalyptic structure of feeling limits how the future can be thought by multiplying ominous signs of a looming disaster at the limits of current comprehension (Aradau and Van Munster 2011). Threatening futures are made present in images of burning cities, talk of crashing economies or graphs predicting imminent environmental ruination (de Goede and Randalls 2009). Life is conditioned in the sense that such structures of feeling may motivate action, or become part of the conditions for action. Future disaster contributes to various forms of ameliorative action or it might become part of reactions that mix boredom and fascination. If Sennett (1998: 31) is right that 'instability is meant to be normal' in precarity, then it coexists with harbingers of total destruction in the form of a sense of the catastrophic or apocalyptic combined with a pervasive, and perhaps pleasurable, ruination.

Linking precarity with the apocalyptic or catastrophic, hints that rather than simply replacing one another structures of feeling may overlap, mutually reinforce one another, blur, become distinct or otherwise relate in complicated ways. Compare precarity with crisis and emergency as ways of governing and experiencing uncertainty. Both terms are used in Western states in relation to a huge array of events and processes: climate change, urban disorder, transspecies epidemics, flooding, eurozone austerity measures and terrorism, to name but some. Emergency and crisis, now used to denote specific situations, share with the feeling of precarity an emphasis on the contingency of life and the uncertainty of the future. Frequently used interchangeably, emergency and crisis are taken to refer to a temporary situation that demands resolution (Anderson and Adey 2012).[21] What they share with precarity is a sense that contingency, and thus uncertainty of some form, is now a normal part of life in liberal democracies. Unlike in relation to precarity, in both crisis and emergency urgent action is demanded and promises to make a difference to mitigate the immediate source of uncertainty. The presumption being that action restores stability from within a situation in which some form of normal order is being overturned. Outcomes are uncertain as normal life is disrupted and disordered. But outcomes are also consequential. Use of the term emergency by governments or other organisations signals that something valued is in danger and that some form of action is demanded. What becomes permanent as both terms get attached to a range of situations is the ever present possibility of a crisis or an emergency. Society is on the verge of disruption. Insecurity and instability constitute a new normal which people are encouraged to learn to live in and with.

What precarity shares with emergency and crisis is uncertainty. Precarity merges with emergency and crisis when the latter become permanent conditions

---

21 Although crisis can also be used slightly differently to refer to a cyclical occurrence that may end in a qualitative shift in the state of a system (as in the periodic crises of capitalism).

of existence and life is supposed to be in or on the verge of disruption. Where precarity differs is that no promises are given that precarity can be ended. Being more specific, the repetition of emergency and crisis is orientated to a space-time that is outside of apocalyptic fantasies of ruination or dreams of progress: the future and present blur into an 'interval' where disaster has not yet happened, action has consequences, and life and death are at stake, or at least are claimed to be. Life is tensed at a threshold in an emergency: between a return to normality, even if it is a normal life always on the verge of another emergency, and a descent into a disaster. Precarity is different: life is suffused with a non-localisable insecurity occasionally interrupted by intensifications around particular futures.

What this brief detour through emergency and crisis reminds us is that, first, structures of feeling resonate with one another without being equivalent and, second, that the boundaries between structures of feeling are indistinct. Structures of feeling may also come to coexist with one another. Precarity occurs alongside what might initially seem to be a contrasting way of dealing with uncertainty and making present the contingency of life and work: the valorisation of flexibility from the 1970s onwards by states and management. Boltanski and Chiapello (2005) have convincingly argued that a sense of 'flexibility' is part of an attempt to mobilise a project of self-realisation in defence of capitalism, where uncertainty becomes a resource to be harnessed for capital accumulation.[22] Related to capacities for commitment and adaptation, the idea and practice of flexibility incorporated a series of critiques of capitalism. 'Flexibility' involved the subordination of the demand for authenticity to a demand for personal liberation in work, especially creativity, freedom and self-fulfilment. Throwing off the constraints of newly old-fashioned social conventions, the structure of feeling of 'flexibility' acts to harness desires for personal liberation to the world of work. Albeit writing somewhat earlier, Donzelot (1991: 251) identified a comparable shift to 'pleasure through work' in which work comes to be seen as 'a means towards self-realization rather than as an opportunity for self-transcendence'. It is against the background of the sense of flexibility as a means of modifying an individual's relation to work and to capital that precarity comes to be identified as an aspect of the contemporary condition: precarity names the pain and costs of perpetual insecurity rather than the freedom 'flexibly' to navigate change and invest hope in (self)transformation.

Juxtaposing flexibility with precarity reminds us that, despite claims, only rarely is an 'age' or an 'epoch' marked by one nameable structure of feeling. Also, a single structure of feeling cannot be reduced to an apparatus (including signifying apparatuses that organise through binary logics and negate otherness).

---

22 Accompanying this increase in 'personal flexibility' has been what Boltanski and Chiapello (2005: 218) term 'internal' and 'external' flexibility. The former refers to changes in the organisation and techniques of work (multitasking, self-control, performance management), the latter refers to the networked organisation of work (the use of subcontractors) and the more malleable character of the labour force (working hours, duration of work and terms of employment).

Apparatuses are not necessarily contemporaneous with structures of feeling. For example, precarity has been understood as a characteristic affective condition of what we could call neo-liberalising[23] and post-Fordist apparatuses. It may be. But it coexists with a range of other affective conditions which may not be smoothly integrated with the apparatuses through which lives are organised and intervened in (as is often assumed in the literature on 'cultures of' fear as an affect/effect of apparatuses of security orientated to threats and dangers, for example). Consider, for a moment, some of the other affective conditions associated with neo-liberal life. We might speculate that neo-liberalising apparatuses work around the pre-emption of any and all alternatives and the generation of affects of inevitability. How might we think about the 'neo-liberal realism' (after Fisher's (2009) capitalist realism) that seems to imbue certain solutions to crisis with a momentum and weight? What is the relation between the certainty of neo-liberal realism and precarity? Closely linked to the orientation to the future in those two structures of feeling, we might consider how Foucault's (2008) brief remarks on 'state-phobia' in *The Birth of Biopolitics* help us understand the critique of the state in neo-liberal reason.[24] 'State-phobia' traverses quite different apparatuses, and changes across those apparatuses. As Foucault (2008: 76) puts it, it has many 'agents and promoters', meaning that it can no longer be localised. It circulates alongside the concern with excessive government, reappears in different places or contexts and therefore overflows any one context (Foucault 2008: 187). Hinting to a genealogy of state-affects, Foucault differentiates it from a similarly 'ambiguous' phobia at the end of the eighteenth century about despotism, as linked to tyranny and arbitrariness (Foucault 2008: 76). Contemporary 'state-phobia' is different. It gives a push to the question of whether government is excessive, and as such infuses and animates neo-liberal policies and programmes. 'State-phobia' is, on this account, both cause and effect of the suspicion that the state is always governing too much. It coexists with 'neo-liberal realism', precarity, and other structures of feeling that mediate neoliberalising apparatuses and are mediated by them.

23 This is to understand neo-liberalism, after Foucault (2008), principally as a critique of the social state (principally the post-Second World War European welfare state) and both a mode of reasoning and set of techniques of government that aim to extend the form of the market to all of life. As with some variants of liberalism, the market becomes simultaneously the limit of and site of verification for government action. Foucault (2008) shows that in neo-liberalism the market is understood in a specific way: it is the market understood in terms of the formal mechanism of competition that becomes the truth and measure of society, rather than the exchange of equivalents between equals.

24 'State-phobia' is described by Foucault (2008: 76) as a secondary 'sign' or 'manifestation' of a crisis of liberal governmentality. As such, it is rarely directly commented upon within the lectures. Foucault (2008) introduces the phrase in the context of a critique of the French Left's relation with the state (76–8), and returns to it in one other lecture (187–91). The passage on 'state-phobia' draws on Anderson (2012).

5.5 Structuring Feeling

My deliberately speculative narrative has moved quickly across precarity, emergency, crisis and state-phobia to draw out how thinking with the concept of structures of feeling presents a series of problems for analysis, specifically how structures of feeling relate to other social-spatial formations and how structures of feeling coexist. What I have attempted to highlight whilst speculating on precarity are some of the ways in which a structure of feeling may coexist with apparatuses, without one being reducible to the other. A structure of feeling will also coexist alongside and may become indistinguishable from or partially connected to other structures of feeling. The result is a kaleidoscopic pattern of pressures and limits, or affective qualities and dispositions, drawing together and crossing diverse domains of life.

How, then, is it possible to understand collective affects such as precarity or fear of mugging or capitalist realism as akin to shared, dispositional, conditions for life? It is this question that has animated this chapter. I have begun to develop a vocabulary specific to the third of the three relations between affect and modes of organisation that the book explores: affect as part of the conditions for the living of life and for the birth of forms of power, as well as an emergent property of encounters and an object-target of apparatuses. One element of that vocabulary is that what is shared when moods are shared may be participation in the same signifying apparatus. The presumption being that a system of signification operates in-between the individual and the social and delimits how events might be felt by individuals. Whilst I have questioned the reduction of all forms of mediation to signification, what this work teaches us is that signifying apparatuses organise patterns of affective relations and investments and, as such, are one way in which collective affects are organised.

The concept of structures of feeling supplements this work by starting from the proposition that collective moods are themselves mediating processes, even if they may 'hover on the edge of semantic availability' (Williams 1977). To 'hover on the edge of semantic availability' is to invite speculation as an appropriate method for discerning and attending to the 'pressures' and 'limits' through which something like a collective mood becomes part of how life is lived and felt and organised. Emphasising the speculative moment of analysis reminds us, however, that articulating a structure of feeling is always a proposition; always contestable. By offering a cultural hypothesis that precarity (as well as state-phobia, flexibility, emergency, and much more) has become central to neo-liberalising and post-Fordist apparatuses, my aim has been to question how structures of feeling emerge and change, form and deform, and perhaps hold steady before then dissipating. On this understanding, differentiated structures of feeling coexist in complex relation with apparatuses and other conditions, some resonating with others, some dominant, some emergent. And a structure of feeling will be one, but only one, mediator within the encounters from which affect as a body's capacity to affect and be affected will emerge and vary. Likewise, the geo-historical formation of a body's 'force of existing' will mediate how a structure of feeling is expressed in feelings, qualified in emotions or otherwise attuned to and sensed. As discussed in Chapter 4, the other elements in an encounter, whether human or non-human, may mediate how a structure of feeling is lived and felt.

Structures of feeling act as affective conditions in the specific sense that they 'bestow an enigmatic coherence' and predispose relations to self, others, and the world. In the following chapter I want to turn to think through a second way in which collective affects mediate and are mediated: the affective atmospheres that emanate and envelop as part of 'lived, practical and unevenly formed and formative experience' (Williams 1977: 89).

Chapter 6
Affective Atmospheres

6.1 Atmospheres

In this chapter I will experiment with the term 'affective atmosphere' as a second response – after structures of feeling – to the problem of how collective affects become conditions that shape without necessarily determining capacities to affect and be affected. If a structure of feeling 'presses' and 'sets limits', how does an atmosphere envelop people and things? How is it possible to attend to the relation between ensembles and something like an atmosphere that may exist ambiguously: 'like a sort of spirit that floats around'(Michel Orsoni 1998 cited in Preston 2008: 7)? In response to these questions I will show that it is the very ambiguity of affective atmospheres – between presence and absence, between subject and object, between subject and subject, and between the definite and indefinite – that enables us to reflect on how something like the affective quality, or tone, of something can condition life by giving sites, episodes or encounters a particular feel.

I am not alone, however, in being intrigued by the notion of affective atmospheres (see Bissell 2010; McCormack 2008). If we understand atmosphere as a term – in Rabinow's (2007) sense of the juxtaposition of a word, a referent object and a concept – then we find it has been used in multiple ways. I can only touch upon some of these here. In everyday English and Anglo-American aesthetic discourse, the word atmosphere is used vaguely and interchangeably with mood, feeling, ambience, tone and other ways of naming collective affects. Each word has a different etymology and specific everyday and specialist uses. Moreover, the referent for the term atmosphere is multiple; epochs, societies, rooms, landscapes, couples, art works, and much more, are all said to possess atmospheres (or be possessed by them). Finally, when atmosphere has been developed into a concept we again find differences. Atmosphere is: impersonal or transpersonal intensity (McCormack 2008; Stewart 2007); environment, or the transmission of the other's feeling (Brennan 2004); qualified aura (Böhme 2006); tone in literature (Ngai 2005); mimetic waves of sentiment (Thrift 2006); or more broadly a sense of place (Rodaway 1994). Of course, we find the same multiplicity when thinking about emotion, affect or any other term that might become part of a vocabulary proper to affective life. This is unsurprising. As I argued in Chapter 1, rather than having been downplayed, repressed or silenced, affective life has been subject to an extraordinary array of explanations and descriptions (Despret 2004). Acknowledging this multiplicity means that we might learn to offer concepts that are equal to the ambiguity of affective conditions, whether those conditions 'press', 'limit' or 'envelop'. Structures of feeling is one such concept, affective atmosphere is another.

My aim in this chapter is not, then, to offer a definitive conception of affective atmospheres. Rather, by holding onto the ambiguities that surround the term I want to learn to sense how things come to have an affective charge. This is to develop further the environmental account of affect offered in the previous chapter. My guides will be two phenomenologists who wonder about atmosphere as an aesthetic concept – Gernot Böhme and Mikel Dufrenne – in dialogue with empirical work on how atmospheres of emergency are crafted in particular sites as the UK government prepare for events. But first to Marx and his material imagination, as a way to introduce some of the problematics that surround the concept of affective atmospheres in relation to structures of feeling.

6.2 The Strange Materiality of Atmospheres

On 14 April 1856, Karl Marx addressed an audience in London on the fourth anniversary of the Chartist *People's Paper*. In a now famous passage, he invokes a certain 'revolutionary atmosphere' of crisis, danger and hope. An atmosphere that floats around, revealing, betraying that:

> The so-called revolutions of 1848 were but poor incidents – small fractures and fissures in the dry crust of European society. However, they denounced the abyss. Beneath the apparently solid surface, they betrayed oceans of liquid matter, only needing expansion to rend into fragments continents of hard rock. Noisily and confusedly they proclaimed the emancipation of the Proletarian, i.e. the secret of the 19th century, and of the revolution of that century … . But, although the atmosphere in which we live, weighs upon every one with a 20,000 lb. force, do you feel it? No more than European society before 1848 felt the revolutionary atmosphere enveloping and pressing it from all sides (Marx 1978: 577).

Marx's metaphorical use of the term 'atmosphere' in this famous address has long interested me. In particular, I have been intrigued by the question Marx addressed to his audience: 'But, although the atmosphere in which we live, weighs upon every one with a 20,000 lb. force, do you feel it?' (Marx 1978: 577). His answer is no. He assumes his audience does not 'feel it', despite it 'enveloping' society from all sides (577). Marx's invocation of the term atmosphere is, of course, part of a complex material imagination that invokes the element of air alongside the state of a fluid ('oceans of liquid matter') and the element of earth ('hard rock'). As such it is thoroughly materialist; albeit a turbulent materialism in which life is imagined through a combination of different elements and different states, in particular solids and liquids (Tiffany 2000; Bennett 2001; Anderson and Wylie 2009). This is, then, a strange materialism appropriate for the reality of affective conditions and far removed from calls to re-ground analysis in the supposed solidity of physical objects or socio-economic conditions.

The revolutionary atmosphere Marx invokes is akin to the meteorological atmosphere in two senses; it exerts a force on those that are surrounded by it, and like the air we breathe it provides the very condition of possibility for life. Marx is not quite invoking an affective atmosphere, even though a revolutionary atmosphere must come charged with a sense of danger and promise, threat and hope. Nevertheless, what intrigued me about Marx's comments when I first read them was how they resonated with the strangle, puzzling, use of the term atmosphere in everyday speech and aesthetic discourse. It is no surprise that a society is taken to possess a certain atmosphere – qualified as 'revolutionary'. As a term in everyday speech, atmosphere traverses distinctions between peoples, things and spaces. It is possible to talk of: a morning atmosphere, the atmosphere of a room before a meeting, the atmosphere of a city, an atmosphere between two or more people, the atmosphere of a street, the atmosphere of an epoch, an atmosphere in a place of worship or the atmosphere that surrounds a person, amongst much else. Perhaps there is nothing that does not have an atmosphere or can become atmospheric. Used in this way the idea of atmosphere blurs with the concept of structures of feeling, even if it lays more emphasis on singularity. It is always the atmosphere of something specific that can be separated from constitutive relations.

Nevertheless, in these brief reflections I think we can also find a first and important difference between a structure of feeling and an atmosphere. A structure of feeling sets 'limits' and exerts 'pressure', even as it hovers on the edge of semantic availability (section 5.3). We could make a comparison with the concept of 'apparatus' as means of thinking through a shared disposition for how life is lived and relations come to be organised. Hence the link between structures of feeling and the concept of mood that I touched on: both bestow an 'enigmatic coherence' across domains. Atmospheres also condition, but in a different way. An atmosphere 'surrounds' and 'envelops' something particular, whilst also existing on the edge of semantic availability. Returning to a 'revolutionary atmosphere', Marx hints to a seemingly counter-intuitive assumption about the reality of atmospheres and how an atmosphere may come to condition life and thought. On the one hand, atmospheres are real phenomena. They 'envelop' and thus press on a society 'from all sides' with a certain force. On the other hand, they are not necessarily sensible phenomena. Marx has to ask if his audience 'feels it'. He assumes not. Nevertheless, atmospheres still effect with a certain force – albeit in a way that may be only tangentially related to any subject who feels.

Of course, Marx's use is metaphorical. But perhaps this use of atmosphere in everyday speech, architectural writings, aesthetic discourse and elsewhere provides a slightly different 'environmental' understanding of affect to that outlined previously. Atmospheres envelop people, things, sites. There is a link back to the concept of affect, if we untie affect from any specific body's capacities to affect and be affected. As well as discussing affects in terms of pre-personal intensities, Deleuze and Guattari (1987) also describe atmospheric phenomena across literary and everyday examples in a way that recalls Marx's turbulent materialist imagination: in the conditions of rain, hail, wind and air favourable to the transport of affects in demonology; Charlotte Brontë's description of love, people and things in terms of wind; the affect of white skies on a hot summer day; or wonder as rainbows form in *Les Météores* by Michel Tournier (Deleuze and Guattari 1987: 288–9).

What do these links between Marx's material imagination, meteors and Deleuze's environmental understanding of affect tell us about affective atmospheres? Apprehending atmospheres through the element of air or the state of a gas reminds us that atmospheres are material phenomena, but their materiality is strange. Perplexingly, the term atmosphere seems to express something vague. Something, an ill-defined indefinite something, that exceeds rational explanation and clear figuration. Yet, the affective qualities that are given to this something are remarkable for how they affirm the singularity of this or that atmosphere and the inseparability of the atmosphere from that which it emanates from. Think of the breadth of qualities used to describe affective atmospheres in everyday speech or aesthetic discourse: serene, homely, strange, stimulating, holy, melancholic, uplifting, depressing, pleasant, moving, inviting, erotic, collegial, open, sublime (Böhme 1993). Atmospheres would, on this account, be close to collective 'vitality affects' as described by Stern (1998: 54): dynamic qualities of feeling such as 'calming', 'relaxing', 'comforting', 'tense', 'heavy', 'light' that animate or dampen the background sense of ongoing life and can be attached to more or less anything. A vitality affect names a 'contour' of experience and as such attempts to express something like the tone of everyday activities. Stern is concerned with subjectivity, and describing affects that cut across the neuropsychological, the bodily, the conscious and interactional, so the comparison does not quite work. Nevertheless, invoking 'vitality affects' does give us a sense of the variety of atmospheres. It also reminds us that atmospheres may involve forms of feeling that may not be captured in the names we have for discrete emotions or feelings and that enveloping and surrounding can be dynamic processes.

By linking the concept of affective atmospheres to a turbulent material imagination we reach a first approximation of atmospheres: the vitality affect of a particular thing. Yet the idea of collective affects conditioning life has been subject to numerous prohibitions, silences and bans amid the many attempts to link affectivity to human species-being (Seigworth 2006). The consequence is that reflections on atmosphere and linked concepts have formed a subterranean current that extends far beyond the recent interest in the work of Massumi, Sedgwick and other affect theorists. From reflections on the panic and hatred of crowds in early twentieth-century crowd psychology (Brennan 2004), through to Maffesoli's (1996) 'affectual tribes', we find long-standing speculations on the nature of something like atmospheres. Most recently, a range of work drawing on the turn to consider the neurology and biology of the body has emphasised how atmospheres are formed through somnambulistic imitation rooted in the body's biochemistry (Brennan 2004). Notwithstanding substantial differences between these ways of encountering atmosphere, all draw out the ambiguities that surround atmosphere, and link terms such as aura, tone or ambience. It is these ambiguities that I discuss in the next section, specifically: atmospheres as finished and unfinished; atmospheres as a property of objects and a property of subjects; and atmospheres as reducible to bodies affecting other bodies and exceeding the bodies they emerge from. This will enable us to explore the specific type of 'envelopment' that an atmosphere is and think through a second type of affective condition that, as with structures of feeling, mediates whilst retaining a tension between the structural and the ephemeral, the formed and the formless, the determinate and the indeterminate, and the subjective and the objective.

6.3 Affective Qualities

To begin to think the relation between atmosphere and the singularity of affective qualities I find inspiration in a seemingly unlikely source – the mid twentieth-century Husserlian phenomenologist Mikel Dufrenne (1976).[1] Dufrenne provides one of the few explicit reflections on the concept of atmosphere in his classic work on the phenomenology of aesthetic experience. Echoing the concern with corporeal experience in phenomenology, Dufrenne's interest was with aesthetic experience in the Greek sense of *aistësis* – 'sense experience'. What I want to draw from Dufrenne's work is the unfinished quality of atmospheres. Atmospheres are always forming and deforming, appearing and disappearing. They are never still, static or at rest. His account of the dynamism of affective atmospheres was developed as part of an attempt to distinguish aesthetic objects from other types of object, where aesthetic objects are a 'coalescence of sensuous elements' (Dufrenne 1976: 13). For Dufrenne, the 'irresistible and magnificent presence' (86) of aesthetic objects establish the conditions for representation to occur. Rather than represent an existing world, a perceived work of art expresses a certain bundle of spatial-temporal relations – what Dufrenne calls an 'expressed world'.

Consider how Dufrenne describes dance as a set of corporeal techniques for crafting 'expressed worlds'; a description that resonates with work in

1 *Seemingly unlikely* because, as Rei Terada (2003: 11) argues, Dufrenne's *The Phenomenology of Aesthetic Experience* is 'a veritable encyclopaedia of the ideologies of emotion to which post-structuralism reacts'. She stresses that Dufrenne (1976) relies on a transcendental theory of emotional expression, where feeling is the expression of a 'primordial unity' linking subject and object, or a 'communion'. The problem is the reciprocal unity between subject and object that is presumed by Dufrenne. What is of interest to me is how the aesthetic work exists as a 'quasi-subject', to use Dufrenne's terms, through its aesthetic quality; a quality that comes to organise the sensuous. It is worth noting though, as Terada does, that Dufrenne is cited positively by Deleuze in *Cinema 1* where Deleuze (1986) draws a link between Pierce's idea of 'firstness' and Dufrenne's material or affective a priori (Chapter 6, n16).

non-representational theories on dance and the creativity and force of expression (see McCormack 2003):

> The atmosphere is what the dancers aim at. It is the aesthetic object itself as they bring it into being. This atmosphere is perceptible even in the pure dance, where expression is not prompted and fortified by a specific subject. The dance always expresses, even when it does not narrate. It is grace, lightness, innocence. In this sense the dance triumphs as beyond all representation, as an absolute language that bespeaks only itself (Dufrenne 1976: 76–7).

Atmosphere is the term Dufrenne uses for how the 'expressed world' overflows the representational content of the aesthetic object as '[a] certain quality which words cannot translate but which communicates itself in arousing a feeling' (178). Throughout *The Phenomenology of Aesthetic Experience* (1976), atmosphere is used interchangeably with other terms – including luminescence of meaning (188), interiority (376), and the unconditioned (194) – as part of a conceptual vocabulary attentive to this other non-representational form of communication. The classic aesthetic 'affective qualities' would be the sublime, tragic, comic or beautiful. But Dufrenne also gives numerous other examples of what, after Ngai (2005), we could call minor atmospheres, including the 'grace, lightness and innocence' of dance (Dufrenne 1976: 76), the 'nobility, fervour, majesty, [and] tranquillity' of architecture (179), the 'indifferent cruelty' of a writer like Zola (178) or 'the lightness of childhood' in Woolf's *The Waves* (183). Take his discussion of scenery and set design:

> Consider a setting like that of Bérard for *Les Bonnes*. Because it is presented as stuffy, sumptuous, and suffocating, an apartment is able to become the principal personage in a play. So, too, is a forest filled with mystery, especially when contrasted with the liberating sea, as Valentine Hugo has demonstrated in the case of *Pelléas et Mélisande*. In such instances the affective quality of the world matters more than its geography. ... The scenery ceases to decorate because it has undertaken the responsibility of expressing the world rather than leaving it to the care of the text (179).

Stuffy, suffocating, nobility, grace, heaviness, and so on, are all names for singular affective qualities that emanate from the aesthetic object as a whole and express worlds. In the passage below Dufrenne describes more formally what an atmosphere is and does in a way that resonates with Orsoni's (1998) description of atmosphere as 'a sort of spirit that floats around' that I cited in the chapter's introduction. Dufrenne also allows us to develop the argument that an atmosphere as the vitality affect of a particular body (where 'body' may be anything from an object to a stage to a person):

> Thus it [atmosphere] is a matter of a certain quality of objects or of beings, but a quality which does not belong to them in their own right because they do not bring it about. The quality in question is like a supervening or impersonal principle in accordance with which we say that there is an electric atmosphere or, as Trenet sang, that there is joy in the air. This principle is embodied in individuals or in things. … Whether or not it is a principle of explanation, it is at any rate a reality that we feel keenly when we come into contact with the group from which it emanates. We have much the same experience in a dark forest. It seems to us that individual shadows are not the result of shade, but, on the contrary, that the shadows create the leafy summits and the entanglement of underbush along with the entire vegetable mass in its damp mystery. The forest prevents us from seeing the tree, and the forest itself is seen only through its atmosphere (168).

Note how Dufrenne does not settle on a clear definition of what an atmosphere is. Instead, he offers a series of approximations in order to attend to the aesthetic object. Atmosphere is 'like' a 'quality of objects or beings', a 'supervening or impersonal principle' or a 'collective consciousness' (here his multiplication of definitions resonates with Williams's definition of 'structures of feeling' – section 5.3). By equivocating about what an atmosphere 'is' Dufrenne expresses that atmospheres may be indistinct, their existence perpetually in question, their reality ambiguous. And yet at the same time subjects and objects are within atmospheres and we encounter particular things, other people or sites through them. The forest in Dufrenne's example is 'seen' through the atmosphere: a quality of shade.

What is common across these approximations is that an atmosphere is a singular affective quality that is irreducible to a series of interacting, component parts. As Dufrenne (327, italics in original) stresses, 'We cannot reduce to their elements the melancholy grace of Ravel's *Pavana pour une enfante défunte*, the glory of Franck's chorales, or the tender sensitivity of Debussy's *La fille aux cheveux de lin*'. Whilst an atmosphere is composed of elements, the aesthetic quality exceeds them. And through this singular affective quality, the aesthetic object creates an intensive space-time. A space-time that exceeds lived or conceived space, even as it 'emanates' from the material and representational elements that compose the art work. Thus:

> The architectural monument has a grandeur or a loftiness incommensurable with its surface or its height. The symphony or the novel has a rhythm, a force, or a restraint of which an objective measure like the metronome gives only an impoverished image. We should realize that, in seeking to grasp expression, we disclose an unpopulated world, one which is only the promise of a world. The space and time which we find there are not structures of an organised world but qualities of an expressed world which is a prelude to knowledge (183).

The atmosphere of an aesthetic object discloses the space-time of an 'expressed world' – it does not represent objective space-time or lived space-time. It creates a space of intensity that overflows a represented world organised into subjects and objects or subjects and other subjects. Instead, it is through an atmosphere that a thing, person or site will be apprehended and will take on a specific presence. Examples abound of this disclosive role in Dufrenne's writings; a feeling of emptiness communicated by a chilling verse, a tragic feeling in Macbeth or the motionless opacity of Cezanne's landscapes, amongst others.

Through his examples, Dufrenne raises the question of how exactly atmospheres – which are themselves ambiguous – 'emanate', or in the passage below are produced:

> This quality proper to a work – to the works of a single creator or to a single style – is a world atmosphere. How is it produced? Through the ensemble from which it emanates. All the elements of the represented world conspire to produce it, according to their mode of representation. ... thus between the situations he is caught and the visages the world offers to him. In another instance, the unity will arise from a certain allure, from a rhythm common to events, from a style of the world creating a style of life and not the converse ... . In still another case, the unity will proceed from the very rhythm of the story, as in the ardent or peaceful breathing of a world which hardens into fatality in the short stories of de Maupassant (136).

What is so important about this passage is that Dufrenne is open about how an atmosphere may 'proceed' or be 'produced'. He does not reduce the production of an atmosphere to a secondary effect of some form of biophysical transmission, as we find in some recent work on atmosphere that attempts to specify some kind of single, identifiable, biologically rooted process in order to explain how atmospheres come to be shared (notably Brennan 2004). Dufrenne gives two examples of how a singular affective quality may be produced from different types of ensemble: from a certain 'allure' and from the 'rhythm of the story'. Obviously, Dufrenne's concerns are aesthetic, indeed literary, in this passage so we need not go into more detail about these particular ways. What is important to note is that there may be a variety of ways through which atmospheres are produced. And yet there is also an assumption that atmospheres cannot be reduced to any one element taken in isolation from the others. I will return to the relation between atmosphere and the 'ensemble' that atmospheres emanate from, but for now we should note that the emphasis on how atmospheres 'proceed' or are 'produced' in relation to ensembles opens up questions of causality, an issue I will also return to in section 6.5.

The intensive space-times expressed through aesthetic objects are not self-enclosed. For Dufrenne, the atmosphere of the aesthetic object elicits a feeling or emotion in a spectator, viewer or listener which 'completes' the aesthetic object and 'surpasses' it (521). The singular affective quality of an aesthetic object is 'open' to being 'apprehended' through bodily capacities and emotions (here 'apprehended' can be thought in terms of expression and qualification).[2] What is interesting about this account, for my purposes, is that atmospheres are unfinished because of their constitutive openness to being expressed and qualified in specific encounters. Atmospheres are indeterminate. They are resources that must be attuned to by bodies. Dufrenne invokes the ineffable when describing atmospheres. He also stresses that atmospheres exceed clear and distinct figuration because they both exist and do not exist. On the one hand, atmospheres require completion by the subjects that 'apprehend' them.[3] They belong to the perceiving subject. On the other hand, atmospheres 'emanate' from the ensemble of elements that make up the aesthetic object. They belong to the aesthetic object. Atmospheres are, on this account, always in the process of emerging and transforming. They are always being taken up and reworked in the events of lived experience: being expressed in feelings and qualified in emotions that may themselves become elements within future atmospheres. But they are not reducible to individual sense experience given the emphasis Dufrenne places on the encounter. This insight is an important one because it reminds us to consider how atmospheres embody a tension between the subjective and objective: atmospheres are 'revealed' by feelings and emotions but are not equivalent to them. In this process, what results is a multiplication of ways in which an atmosphere is translated into individualised experience during encounters. Perhaps what Dufrenne says about the aesthetic object also holds for atmospheres, providing we understood 'interpretations' broadly: '[T]he aesthetic object gains being from the plurality of interpretations which attach themselves to it' (65). Atmospheres, emanating and enveloping particular things, sites or people, are endlessly being formed and reformed through encounters as they are attuned to and become part of life.

Whilst I remain cautious about aspects of his account, and Dufrenne's arguments do concern the distinctiveness of the encounter with aesthetic objects, his work does help us think of atmospheres in terms of singular 'undifferentiated' affective qualities that express a world and are not simply reducible to one particular biological or cultural mechanism of affect transmission and circulation.[4]

---

2 Dufrenne's account of the 'apprehension' of atmospheres is specific to aesthetic objects that he distinguishes from everyday objects (a distinction that perhaps no longer holds, if it ever did, given the anaesthetisation of everyday objects). Involved or enveloped in the art work itself, everyday reality is suspended for the viewer or spectator: '[I]t is in the work that they find meaning, not in themselves as something to transfer onto the work. … But we should add at once that it is the work which awakens us to ourselves' (1976: 59–60).

3 Dufrenne describes the encounter and thereafter relation between the aesthetic object and spectator in the following terms: '[T]he reality of this object can only be revealed, not demonstrated. It has no other guarantee than to be attested to by a perception and to be situated at the crossroads of a plurality of perceptions' (1976: 47).

4 Specifically, there are two potential problems in light of my other arguments in this book. First, Dufrenne's emphasis on a 'communion' between subject and object rather than a differentiated set of relations. Second, his concept of the 'affective a priori' which can be read as grounding experience in a structure that is outside of history.

His work invites us to think through how affective qualities emanate from ensembles and how they are constitutively open to being differently expressed in affects and differently qualified in named emotions. Dufrenne's emphasis is firmly on the affective quality of aesthetic objects, and his examples all concern the realm of what we might term 'high art' and emphasise the difference between the aesthetic object and the ordinary doings of everyday life. However, it is not clear why we should restrict the production of singular affective qualities to sculpture, music, architecture or other seemingly self-enclosed aesthetic works. Numerous bodies can be said to be atmospheric, in the sense that people, sites or things produce singular affective qualities and emanate something like a 'characteristic' or a 'quality'. This expansion of affective atmosphere to multiple bodies is the starting point of Gernot Böhme's (1993; 1995; 2006) ecological aesthetics. It is risky. The risk is that it ignores how an atmosphere 'gains being' from a plurality of encounters, to paraphrase Dufrenne. Yet it is worth exploring because it enables us to think further about the spatialities of atmospheres – how atmos-*spheres* condition by enveloping and surrounding.

6.4 Atmos-*spheres*

Böhme's basic definition of atmospheres shares much with Dufrenne but he puts more emphasis on the spatiality of atmospheres, describing them as 'spatially discharged, quasi-objective feelings', 'tuned spaces' (Böhme 2006: 16) and akin to 'indeterminate spatially extended quality of feeling' (Böhme 1993: 117–18). Böhme notes the ambiguous status of atmospheres with regard to presence and absence and formed and forming, but lays more stress than Dufrenne on their in-between status with regard to the subject/object distinction. Atmospheres are both objective and subjective at the same time, meaning that they are both and neither:

> atmospheres are neither something objective, that is, qualities possessed by things, and yet they are something thinglike, belonging to the thing in that things articulate their presence through qualities – conceived as ecstasies. Nor are atmospheres something subjective, for example, determinations of a psychic state. And yet they are subjectlike, belonging to subjects in that they are sensed in bodily presence by human beings and this sensing is at the same time a bodily state of being of subjects in space (Böhme 1993: 122).

Böhme does risk re-inscribing a subject-object division by postulating atmospheres as taking place in-between the subject and the object. Nevertheless, he holds the subjective and objective in tension, resisting the urge to bring atmospheres down on one side of the divide. Additionally, he does not simply erase differences between subjects and objects by vaguely claiming that atmospheres are simply impersonal intensities unmoored from actual bodies.

If we compare with another important discussion of the concept we can see why this move is so important. Ahmed (2007/8: 2010) has recently argued against what she refers to as an 'outside in' model of thinking the relation between atmosphere and subjectivity (in part through a discussion of Brennan (2004)). As she sees it, discussion of atmosphere can presume that bodies are neutral when they encounter, read or otherwise relate to atmospheres (and this is certainly a problem in Dufrenne and his classical understanding of a unified community of feeling despite his emphasis on 'interpretation'). Responding to the opening sentence of Brennan (2004: 1) 'Is there anyone who has not, at least once, walked into the room and "felt the atmosphere"', Ahmed rightly sounds a note of caution. Invoking an occasion when an atmosphere might have been misrecognised, she draws out what she sees as a problem with the concept of atmosphere:

> So we may walk into the room and "feel the atmosphere", but what we may feel depends on the angle of our arrival. Or we might say that the atmosphere is already angled; it is always felt from a specific point. The pedagogic encounter is full of angles. How many times have I read students as interested or bored, such that the atmosphere seemed one of interest or boredom (and even felt myself to be interesting or boring) only to find students recall the event quite differently! Having read the atmosphere in a certain way, one can become tense: which in turn affects what happens, how things move along. The moods we arrive with do affect what happens: which is not to say we always keep our moods (Ahmed 2007–2008: 126; see also Ahmed (2004: 10) for a similar point).

Ahmed's intervention is useful because she avoids assuming that atmospheres are simply 'received' by a neutral body (turning us back to the issue of the geo-historicity of a body's 'force of existing' as discussed in Chapter 4). She reminds us again of the dangers of hypostasising 'atmosphere' and the need to focus on specific processes of agency and causality (a task I turn to below). Instead, we get a more complicated picture of the dynamism of atmospheres as they resonate, interrupt, interfere with, change, challenge, disrupt and otherwise intersect with other elements within encounters and a body's existing 'capacity to affect and be affected' (an understanding consistent with Dufrenne's claim that atmospheres 'gain being' from a plurality of 'interpretations'). Nevertheless, it is worth noting the equivocal status given to 'atmosphere' by Ahmed in the above passage. Her use of the terms 'read' and 'felt from' appear to presume that atmospheres are real phenomena that pre-exist encounters. Alternatively, Ahmed might be conceptualising atmosphere as an effect of processes of reading. Indeed the phrase 'the atmosphere is always angled' would suggest that this is the case. Ahmed's equivocation on this point can be understood as a variant of Dufrenne's and Böhme's ambiguities. The concept of atmosphere does not fit into either side of a subjective/objective distinction or terms such as the impersonal that have been experimented with to overcome the divide. We find ourselves 'in' atmospheres that pre-exist us but, as with Dufrenne, atmospheres 'gain reality' through processes of attunement and change through attunement. An atmosphere's openness to change as it is emanated, expressed and qualified makes it less a property, a finished thing in itself, and more a condition constantly being taken up in experience. We are in the midst of atmospheres that constantly flip between the objective and subjective, undoing the distinction between the two terms.

Following on, and despite the connection between atmospheres and the materiality of sites, atmospheres are not unproblematically locatable. Returning to Böhme (1993), he stresses that atmospheres are ambiguous with regard to their location. It is difficult to say where an atmosphere is since '[t]hey seem to fill the space with a certain tone of feeling like a haze' (113–14) even as 'we sense what kind of space surrounds us. We sense its atmosphere' (Böhme 2006: 402). Filling and surrounding; the term atmosphere reminds us of situatedness and immersion within an environment rather than a hyperactive world of endless flows and relations. Emphasising their environmental character, Böhme draws on the German phenomenologist Alfred Schmitz to introduce the strange spatio-temporalities of atmospheres, at least when seen from the perspective of lines and flows:

> Atmospheres are always spatially "without borders, disseminated and yet without place that is, not localizable" (Schmitz). They are affective powers of feeling, spatial bearers of moods (Böhme 1993: 118–19).

Here Böhme returns us to the materialist roots of the word atmosphere touched on previously: *atmos* to indicate a tendency for qualities of feeling to fill volume like a gas, and *sphere* to indicate a particular form of spatial organisation based on the circle. Together they enable us to think how atmospheres are connected to particular 'envelopments' that surround people, things and environments. On this understanding, the spheres expressed by and created by atmospheres would vary, perhaps dramatically. Following Sloterdijk (2005), we can think of a plurality of partially connected/disconnected spheres involving different forms of being-together. This gives a vision of a world of spheres, in which weather, gardens, rooms, people and any other type of being-together may be caught up in atmospheres perpetually forming and deforming.

> Thus one speaks of the serene atmosphere of a spring morning or the homely atmosphere of a garden. On entering a room once can feel oneself enveloped by a friendly atmosphere or caught up in a tense atmosphere. We can say of a person that s/he radiates an atmosphere which implies respect, of a man or a woman that an erotic atmosphere surrounds them (Böhme 1993: 113–14).

There are two different 'envelopments' being hinted at by Böhme in this passage. The first – and most general – is the spatiality of the 'sphere' in the sense of a certain type of surround. Note how an atmosphere 'surrounds' a couple or one finds oneself 'enveloped' by an atmosphere. The centre and circumference of an affective atmosphere may, however, be indefinite or unstable; especially if an atmosphere is taken not only to occupy a space but to permeate it. Thus affective atmospheres may 'leak out', overflowing ways of bounding a sphere. An atmosphere within a group may be felt by another group, for example (Bissell 2010). Atmospheres within a specific site may come into contact with one another, changing as they do. The second spatiality is again spherical but it is, more specifically, a dyadic space of resonance – atmospheres 'radiate' from an individual to another. They appear and disappear alongside the dynamics of what Sloterdijk (2005) terms 'being-a-pair'. This process does not necessarily have to occur between two people. As we have seen with Dufrenne, a space of resonance may be created between an artwork and a viewer who encounters that work, or perhaps the interior of a building as encountered habitually by someone walking through it may emanate an atmosphere. In both cases we find that atmospheres are interlinked with forms of enclosure – the couple, the room, the garden – and forms of circulation – enveloping, surrounding and radiating. Atmospheres have, then, a characteristic spatial form: diffusion within a sphere.

If the characteristic spatiality of atmosphere is the sphere, and atmosphere is the tension between subjective and objective, then we need some way of understanding the composition of atmospheres. Returning to Dufrenne and Böhme, they show us that atmospheres are generated by bodies – of multiple types – affecting one another as some form of envelopment is produced. Atmospheres do not float free from the bodies that come together and apart to compose situations. In addition, atmospheres may be shifted, intentionally or otherwise. The question of how atmospheres are produced is further complicated if we consider the heterogeneity of ensembles. We need to ask how atmospheres are mediated by specific elements within an ensemble, or a constellation of those elements. Here it is important to rethink ideas of causality to understand how atmospheres mediate/are mediated as the affective quality of a particular thing that envelops those who encounter it[5]. It is to an understanding of causality that I now briefly turn by means of an event, or what might appear to be an event, or what came to be felt to me and perhaps others as an event.

6.5 Atmospheres and Causality

> We wait. The small sub-group in a side room from the main room knows that something is coming. The exercise tests a "major incident", so a "major incident" must be on the way. There has to be an event. But when and what will it be? A

5 Beyond the focus here on atmospheres, rethinking causality has been central to the 'affective turn' more broadly (and linked approaches such as performance and performativity). As Hardt (2007: viii) notes 'Affects require us, as the term suggests, to enter the realm of causality, but they offer a complex view of causality because the affects belong simultaneously to both sides of the causal relationship'.

> facilitator introduces a series of occurrences: a car crash, racist graffiti, reports of an explosion in an apparently disused house. Seemingly random, everyone knows they are not the main event. Something else is coming. But what's the pattern between them? And which one is designed to confuse us? Puzzled, no one is quite sure what they add up to. Suspicious, the "moderator" is questioned by people. At times, knowing looks are exchanged. This goes on for five minutes or so. We continue to wait. Suddenly, radios crackle into life. We are called back into the main room. Laughing, it's clear that we've missed what the event is, but excitement grips people. Folders are stuffed into bags, maps and pens swept up, as wordlessly we hurry into the main exercise room. "It's starting then", someone calls out (Author's Fieldnotes from Exercise Observation, 5 June 2010).[6]

I am in an emergency services control room, observing an exercise that practised response to a chemical, biological, radiological or nuclear event. We start off in one of the side rooms, before moving to the main room. Excitement, tension, urgency, anticipation; these and other words could be used to name something close to the atmosphere of the exercise, or was it a set of atmospheres? Waiting for an event to happen, something is shared between us. Perhaps.

Eight minutes later, something happens. Things change.

> A "situation report" is handed out. Read it, instructs the moderator. People do. In contrast to when people entered the room and greeted one another, no one speaks. On one wall of the room images of explosions have been projected. "What would your organisation be doing now?", asks the moderator. Groups of two or three people are talking. Occasional laughter mixes with the muffled sounds of discussion. Stray words interrupt the silence. Sat around a U-shaped table, each little group has moved their chair closer together. Two minutes to go. Sheets are wordlessly filled in. One minute to go. People read through their notes. They have been asked by a moderator to specify the actions their organisations would take in the immediate aftermath of the event and know

6 An exercise is a simulation of an emergency event in which those tasked with response, including the emergency services, rehearse response 'as if' the event was real (see Anderson and Adey 2012). The exercise was observed by myself as part of a two-year (2009–2011) Economic and Social Research Council-funded research project on exercises and emergency planning, conducted with Prof Peter Adey. Practically, the observations involved attending an exercise from start to finish, talking informally to exercise participants and observing the exercise 'play'. The observations focused on; how exercises were organised by emergency planners; the particular types of actions involved in exercise play; the dynamics of exercises, including their atmospheres; and the relationships between different organisations. Any identifying details about the location of the exercise discussed in this chapter have been removed at the request of the organisations involved.

> that those actions will soon be subjected to discussion and criticism (Author's Fieldnotes from Exercise Observation, 5 June 2010).

Perhaps we could name the affective qualities that enveloped the exercise in the control room as anticipation, urgency and focus, whilst noting that exercises are made through many other, additional, atmospheres: pressure, fun, and so on. Neither categorical emotions, nor free floating intensities, atmospheres name the feeling of participation in an exercise as it takes place as an event at a particular site. Taken together, urgency, anticipation and other atmospheres name something close to the characteristic feel of the exercise as a particular type of occurrence. An affective quality that, for a limited time, envelops participants in an 'as if' world – an 'as if' world that is now central to an apparatus of emergency planning.[7]

A word spoken at the right moment, an image that calls to be attended to, a document that needs to be referred to, the presence of an individual, a future possibility: a variety of things act as atmospheres are made and unmade. For Dufrenne (1976: 136), atmosphere is 'produced' 'through the ensemble from which it emanates'. Likewise, Böhme (1995: 33) opens up room for a materialist analysis of how atmospheres issue forth, how they emanate. Atmospheres are '[s]paces "tinted" by the presence of objects, humans or environmental constellations … . They are spheres of presence of something, of its reality in space'. Thus atmosphere is:

> Not something that exists by itself in a vacuum, but quite the opposite. It is something that emanates and is created by things, by people, and by the constellations that happen between them (Böhme 1995: 33–4).

The use of the term 'emanate' by both Dufrenne and Böhme suggests that atmospheres 'issue forth' from particular ensembles, acting like a gas in that atmospheres do not have fixed borders or boundaries yet come to envelop participants. Although the term 'emanate' has phenomenological roots, and as an everyday word offers a useful sense of something moving forth from a source, it does not tell us much about the specific ways in which atmospheres are constructed, produced or crafted. In the above scene, we might note, in no particular order, the following elements that cohere into an ensemble that possesses something like a 'quality of objects or beings', a 'supervening or impersonal principle' or a 'collective consciousness', to use Dufrenne's terms. Noting, as we disentangle the ensemble, that the exercise itself is but one part within an apparatus of security –

7 Whilst not our concern here, exercises are a necessary part of the apparatus of civil protection. UK emergency planning and response is set up around networks of planning and event-specific networks that are primed to be actualised in the event of an event. What an exercise does, or what it should do, is allow for rehearsal of how the event-specific network will meet an event (see Anderson and Adey 2012).

UK emergency planning and response – organised around the urgent need of grasping and handling events:

A smile
Laughter
Tea and coffee
A moderator walking around
The spectre of a chemical, biological, radiological, nuclear event.
Audit and its cultures
A question asked
Organisational protocols and procedures

Far from exhaustive, what such an open-ended list recognises is the diversity of causally efficacious elements that come together as space is 'tuned'. Elements such as a future threat that, as we shall see, appear to be made present through their own atmosphere. And elements that include, we must note, a whole series of ways in which the spatial and temporal boundaries of the exercise at a distinct site are established and endure. From this starting point we get a sense of how atmospheres are emergent properties of an ensemble of diverse elements that hold together, if only for a finite period. A few minutes later we could add new elements to the list: the expectation as the clock ticks; the comfort provided by the seat; the whirl of the power-point projector; light in eyes. However named, the characteristic affective qualities of an exercise are made through the contingent ensemble formed through these mundane bits and pieces. An affective quality comes to envelop and encircle participants, whilst the particular elements may emanate a series of minor atmospheres. Friendship surfacing as colleagues meet and greet one another, boredom beginning to settle as subsection 3b of section 4 of 'Organisational Protocol 7' is flicked through.

To speculate on how an atmosphere forms it is first necessary to understand how a diverse grouping of things and people come together. This is well known by those arts and sciences that aim to shape and manipulate atmospheres, albeit often phrased in ways specific to the practices, techniques and events that make up different practical arts. Indeed, it is precisely the circumvention and emanation of atmospheres that are acted upon when atmospheres become the object-target of explication and intervention in apparatuses. Think of how atmospheres are sealed off through protective measures such as crime prevention and environmental design, or how atmospheres are intensified by creating patterns of affective imitation in sports stadia and concert halls (Borch 2008). Of particular importance in the first scene from the exercise is the situational report[8] that is handed around and acts as

8 A situation report is a device that summarises for a responding agency what is happening during an event at a particular moment of time. Situation reports can be verbal or written and are normally from one agency. Within the apparatus of UK civil contingencies, they have two functions. First, to render the event actionable as it unfolds. Second, to

one amongst a number of ways of staging the event. Rendering that event present in representational form, the situational report acts to stage the chemical, biological, radiological or nuclear event as a specific occurrence happening in a defined locale. The report does not act alone. The material space of the exercise is organised so that players might participate in an event of response 'as if' it were real. Moderators attempt to ensure the suspension of disbelief by, for example, playing a loop of past response on a screen at the front of the room. Taken together, these different materials combine to prime specific atmospheres: the quiet focus that settled over the group as conversations faded; the sense of urgency as people prepared to respond; the anticipation of what was to come. For members of the emergency services accustomed to responding in a particular way, the atmosphere of an exercise has become a familiar part of ordinary work routines. People come to an exercise with an expectation of how to act and memories of past response layered into their acts and habits, leading to expectation tinged with a feeling of having been here before. We could call this a 'force of existing' specific to those who act in response to emergencies and prepare for emergency response: the atmosphere 'gains being' from the plurality of those common dispositions and habits, including my own as I slowly became more comfortable as I observe exercises.

Dufrenne and Böhme invite us to wonder about how atmospheres 'emanate', but if we turn to the set of atmospheres that perhaps inhabit the exercise it becomes difficult, if not impossible, to separate out causes from effects. The words and presence of the moderator intensify as he picks up on the mood of the group, for example. The accumulation of the exercise players' small acts of focus and concentration heighten the sense of seriousness that surrounds the event to come. Of course, linear causation may be useful for thinking about some of the ways in which atmospheres emanate, but an attempt to separate out the assembling of atmospheres into effects and determinants is likely to fall short precisely because atmospheres envelop, they infuse and mix with other elements. For example, in the initial scene of a group of people reading – and its attendant atmospheres of urgency and focus – how can we separate out a cause from the heterogeneous actants that all participate, to varying degrees, in the exercise? It becomes impossible to pin a cause to one element within the ensemble from which an atmosphere emanates and in relation to which atmospheres are enhanced, transformed or intensified. Both Dufrenne and Böhme recognise and work with this ambiguity. It is in the background when they stress that an atmosphere cannot be 'decomposed', that atmospheres are irreducible to their parts and insist as 'overall effects'.

Describing the composition of atmospheres through a more complex form of causality is a task shared with some tendencies in practical arts such as video-game design or architecture that have as their task the composition of atmospheres. We

coordinate action in relation to the unfolding event. A situation report acts, then, as a way of, first, knowing the event and, second, gathering together multiple agencies around a shared diagnosis of what future actions are required. In an exercise, a situation report is designed to summarise the 'as if' event in the absence of the actual event.

can think of architecture, for example, as a way of producing specific atmospheric effects and as a means of creating arrangements that allow certain emanations (Preston 2008). Consider the architect Peter Zumthor's (2006: 2) reflection in *Atmospheres* on how he creates a specific type of affective effect, defined by him as 'this singular density and mood, this feeling of presence, well-being, harmony, beauty … under whose spell I experience what I otherwise would not experience in precisely this way'. Drawing on examples from his own practice, he outlines nine ways of creating that atmospheric effect, including through sound, light, temperature and objects operating within a context. Materials reinforce and strengthen one another producing a type of sympathetic coordination between elements and a type of 'total effect' that cannot be decomposed – which Dufrenne and Böhme both attribute to atmospheres as the affective quality of a particular thing. Following Bennett (2010: 33), herself drawing on Hannah Arendt's distinction between origin and cause, we could call this heterogeneous gathering of materials the origin of an atmosphere: in the sense of 'a complex, mobile, and heteronomous enjoiner of forces' that mediates how an atmosphere emanates. The origin of an atmosphere may, of course, be made up of other atmospheres (Bissell 2010).

Urgency and focus are but one element within exercises which are themselves but one element within the apparatus of 'UK Civil Contingencies'. Exercises involve making the future *affectively* present in a way that enables future response to be rehearsed and disruptive events of multiple types to be prepared for. Whilst the exercise is made up of multiple parts – including fleeting atmospheres – it is also itself one part in a particular post-Cold War apparatus of security. Likewise, the different parts that make up the atmosphere of an exercise are simultaneously parts in, for example, the apparatuses of audit whereby 'preparedness' must be measured and quantified or a shift from forms of precaution or pre-emption to modes of resilience. This is one reason for the volubility of atmospheres: the complex origins that atmospheres emanate from will involve elements from other, partially connected, ensembles and apparatuses. The envelopments reflected, expressed and enacted through atmospheres are fragile; the exercise's urgency having to be maintained by visual materials that resonate with a participant's past response to emergencies, for example. The atmosphere of danger that in the War on Terror has become attached to the threat of a chemical, biological, radiological, nuclear event ensuring that attention holds, for a while.

What my comments hint to, then, is a need to articulate modes of causality that are less linear and less focused on separating out cause from context. However, the linked risk is that the specific powers and effectivity of different actors are flattened, in a way that tells us nothing specific about how atmospheres emanate and envelop. The result might be an acausal account that focuses on the 'mutual constitution' of atmospheres.[9] In working towards alternative models of causality,

9 See Connolly (2004) for a critique of both 'efficient causality' and 'acausal' modes of description and explanation – what he elsewhere terms 'the webs of interdefinition so dear to narrative theory' (Connolly 2005: 111).

we might remember that although perhaps rare there will be examples of what De Landa (2005), Connolly (2011) and others term 'efficient causality': where an effect follows from a determinant and that effect proceeds in a linear fashion on a set trajectory. Indeed Dufrenne and Böhme's descriptions sometimes assume such a model by focusing on the relation between an 'overall effect' (whether in relation to an aesthetic object or a building) and an individual body, albeit one with capacities to affect and be affected. But these occasions are rare. Other modes of causality might be at work as atmospheres form and deform, becoming part of the conditions for encounters before intensifying to emerge as objects of thought. It is at this point that atmospheres come to be named: an act that may itself intensify, enhance or otherwise change an atmosphere as intentions, ideas and beliefs are layered into it.

Suddenly, the atmosphere of the exercise changes, or so I think:

> What would you do? What would you do? Twice the question is asked. A pause of four seconds, before the question is repeated. Raising his voice, the member of a government security agency slows his speech and carefully enunciates each word: WHAT ... WOULD ... YOU ... DO? I feel anxious. I look away. It feels unfair. Even though I'm only observing, I'm desperate to avoid becoming the object of questioning. The question is being directed to a chief executive of a local authority who is playing in the exercise. The correct answer is contained within the 'major incident' plan that is being tested in the exercise. Perhaps, he hasn't read it. Perhaps, he has just frozen under questioning. Hesitant, he stumbles over the answer. He tries again. His nervousness catches on. I exchange a glance with my neighbour. Other people's eyes are down, scanning through the plan that has now attained a new level of significance. Suddenly, the questioner breaks off contact with the chief executive. Nervous excitement mounts, as he clicks the power-point through to the next image: the aftermath of an explosion. Suddenly, he has the room's attention and nervousness suddenly becomes something else. Dead bodies mingle with tangled machinery. Poised, the players in the room wait for the next question. Held in suspense, they know that someone will be asked a question (Author's Fieldnotes from Exercise Observation, 5 June 2010).

In this scene the nervous excitement is itself a participant in the exercise: through excitement responders are readied to face and respond to the event. Indeed, it is through the exercise's atmosphere that an equivalence[10] might be established between response as played in an exercise and response as it might actually occur. What is assumed to be shared between actual and rehearsed response is the affective environment in which response to an event will have to occur: in

10 The equivalence between response to an event as exercised and actual response to an event is a sensuous one: it is felt. Exercises can be understood as now mundane technologies for creating affective equivalences between the space of an exercise and future events.

the above moment it is a shared affective quality of pressure from within which decisions and coordinated action must happen, at other stages of the exercise the affective equivalence is a sense of alertness to what the event might do next, or calmness in the midst of almost overwhelming surprise.

Before turning to the question of how we might think about the agency of the different elements that form atmospheres, we could pause and consider how atmospheres may have a form of what Connolly (2011) terms 'emergent causality' as a particular process of affective mediation. Connolly describes 'emergent causality' in the following terms which give a good sense of how something like an atmosphere is both an effect of a gathering of elements and a mediating force that actively changes the gathering it emanates from. The causal powers of an atmosphere is only, however, revealed through those changes:

> Emergent causality is *causal* – rather than reducible to a mere web of definitional relations – in that a movement in one force-field helps to induce changes in others. But it is also *emergent* in that: first, some of the turbulence introduced into the second field is not always knowable in detail in itself before it arrives darkly through the effects that emerge; [and] second, the new forces may become *infused* to some degree into the very organisation of the emergent phenomenon so that the causal factor is not entirely separate from the latter field (Connolly 2011: 171).

Whilst not discussing atmospheres directly, what Connolly provokes us to think about is how an atmosphere is at once an effect that emanates from a gathering, and a cause that may itself have some degree of agentic capacity. Heads bow, people look down and focus returns; all causes and effects of nervous excitement. It is an 'emergent cause' because we cannot be sure of the character of the atmosphere before registering its effects in what bodies do – an atmosphere is revealed precisely as it is expressed in bodily feelings, and qualified in emotions and other actions. This point loops back round to Dufrenne and Böhme's ambiguities, but adds another. As well as being ambiguous with regard to the absence/presence and subjective/objective distinctions, atmospheres are ambiguous with regard to the distinction between causes and effects. It is in this sense that atmospheres come to act *ambiguously* within apparatuses or encounters. They become one causally efficacious element amongst others, but in a way that is uncertain precisely because of their ambiguous status as surrounds that envelop and encircle. If structures of feeling set 'limits' and 'pressure' how apparatuses and encounters form, then atmospheres are perhaps better approached as affective propositions, unfinished lures to feeling a situation, site, person or thing in a particular way that may come to condition encounters. Action occurs from within atmospheres that may be distractedly dwelt within, intentionally shaped, barely noticed or felt as overwhelming presence.

If atmospheres as complex wholes can be said to have an agentic capacity – albeit a curious, ambivalent, agency in which the very existence of an atmosphere may be revealed retrospectively by its effects – then rethinking causality also allows us to understand how particular elements can act to effect how spaces are 'tuned' (Böhme 1993). In the above scenes of something like an event we get a sense of the life of atmospheres as diverse parts enter in and out of relation: a facilitator shouts, a power-point slide changes, the sun warms, conversation ends, chairs angle bodies, heads turn. As relations change and endure, atmospheres form and deform. Not all of the above elements result in changes in the affective atmosphere. Whilst always finite, atmospheres may endure even as the ensemble changes. Alternatively, an atmosphere may form and fade in a moment as two people clash after one of them shouts a repeated phrase that intensifies: 'What would you do?' 'What would you do?' How, then, do specific parts or elements have causal effects and how do atmospheres have various durations, settling for a moment or staying a while, enveloping before dissipating? Furthermore, we might wonder about some of the ways in which an atmosphere 'gets into' a body's 'force of existing' as atmospheres resonate with bodily capacities. In the above scene, my own nervousness about being an academic observing the exercise intensified in my encounter with the local authority officer's display of nervousness. I hoped the question would not be directed at me. I sat into my seat, my eyes averted, my head angled away. Relief as the moderator moves on. Relief as I feel that the tone of his words change. Relief that I was in the background.

Let us return to the above scene from the exercise with these remarks about the causal powers of atmospheres as affective qualities that emanate and envelop in mind. At that point in the exercise, the space was overdetermined by the presence of the moderator and the force of his words and gaze. Linking to a long and diverse tradition of thinking of leadership as a type of affect work, we could say that the force of his repeated question, a question that increased in intensity, acts to catalyse a subtle but palpable shift in the atmosphere of the exercise. Nervousness was crystallised in the sense that a host of elements in the background to any exercise – fear of failure, been watched in public, differences between organisations – were drawn together around the focal point of the repeated question. His question affected. Other tendencies and latencies up until then in the background of the exercise concentrated and intensified, crossing over into the coordinated actions of looking down, turning away, and avoiding the accusing gaze of the moderator and the stuttering of the floundering player. Perhaps in this context the refrain 'What would you do?' – as example of what Riley (2005) calls the 'forcible affect of language' – acts as a catalyst in the midst of the complex 'origin' of multiple elements. The term 'catalyst' is taken from De Landa's (2005) discussion of how to understand complex forms of causality. He uses it to describe how a vast array of entities may become causes without their effect – or even if there will be an effect – being predetermined. In the exercise, an image of an explosion and the absence of words catalyse a change in atmosphere in the sense that a new atmosphere – perhaps without a name until it is retrospectively revealed in effect – was briefly formed and filled the space of the exercise. Surprisingly, given that they are phenomenologists, Dufrenne (1976) and Böhme (1993) attune to something similar. Both show how minor changes, such as a missed note or an out of place word, can cause atmospheres to fall apart or fade away or otherwise change. Atmospheres may slowly dissipate or may be abruptly brought to an end. The atmosphere of a repeated question is just one such way in which atmospheres may change.

Now let us turn to another scene to deepen this sense of multiple causal factors. The scene happens 23 minutes later in the same exercise. Again, the moderator finishes speaking. This time there is a different reaction: reluctance.

> In quick, staccato, words the chief fire officer is summarising how his service will be using the "tactical" response plan. Designed as a "flexible" resource for responders, copies of the plan sit in front of each of the players, some have been flicked through, some have been ignored. The players are tired. They look and act distracted; checking mobile phones, looking at watches. It's near the end of a long day and the event now appears as though it is under control. Suddenly, an image flashes up. A change in the scenario has been introduced by the exercise team. The event has been escalated: a second device has been located in a different part of the city. Details flash up on the power-point at the front of the room. Heads turn to watch. The players are called back to attention. But something is wrong. Some grumble that the "update" seems unrealistic: "we would have checked there", they whisper. Others are just bored. A couple of people go through the motions of flicking through the tactical plan. The moderator changes tone, perhaps realising that some people aren't focused. Five minutes to decide what to do, he shouts. Initially, a hush descends over the group. But then scepticism returns in bursts of conversation, mutterings and requests for clarification (Author's Fieldnotes from Exercise Observation, 5 June 2010).

The change in scenario took place in the midst of a range of atmospheres. No longer enlivened by the affective presence of future response, boredom had settled over the exercise. Players had lost interest as the event had moved from being charged with uncertainty to becoming a normal occurrence that could easily be handled through habitual practices of response and systematic organisational routines. What this means is that the affective presence of the future event changed as it was expressed and qualified in boredom: time stilled, space slowed as the sense of alertness became something else (Anderson 2004b). Lacking the affective quality of plausibility, the newly-introduced scenario was greeted with scepticism. Responders did not believe in it. Disbelief was not suspended. The dynamics of the exercise event are interrupted as the exercise fragments. No longer held together around the future event as a shared object of concern that demands immediate action and was made present through urgency, the exercise's atmospheres shift and fragment into numerous affects and emotions. What initially seemed to be one atmosphere – of urgency when faced with the event – fragmented into multiple conversations, glances, shakes of the head and other actions. Perhaps the moderator's words miss, failing to be enveloped by a sense of plausibility, failing to resonate.

What this scene offers is a sense of how the 'total effect' or 'unity' that Dufrenne and Böhme stress is only ever a precarious achievement. In the exercise a range of atmospheres form and deform, almost coming into conflict. Slowly an atmosphere of irritation begins to intensify between the exercise players, whilst the pressure animated by the presence of an event abruptly ends. People are called back to their proper roles by the force of the event as a disruption that shatters and disturbs. Repaired, although perhaps infused with a certain weariness and perhaps resentment, the exercise never returns to the pressure and urgency that had, hitherto, marked it out as an event that rehearsed response 'as if' real. What we get a sense of, then, is the apparatus through which a certain sense of the event is synthesised so that something like a defined set of atmospheres endures. In the case of the exercise discussed here, we can see how it is composed through a series of generic techniques, including scenarios, used in conventional ways. Scenarios are used as pragmatic devices to organise attention, as are images of events. The moderator's function is to manage player involvement in the exercise. Many of the elements that compose and catalyse atmospheres will reoccur across exercises. For this reason, we can think of repetition of atmospheres in terms of the reoccurrence of elements and specific ensembles of multiple elements. Of course, this is what is done in apparatuses that attempt to produce managed environments. Careful arrangements of materials are a means to condition, without ever fully determining, specific atmospheric effects. Think of the shop window as an attempt to create atmospheric effects through the arrangement of light, clothes and plastic, for example, or how the setting designed into a video game gives the game a specific atmospheric quality through the intensity of light or level of detail (Ash 2010; Miller 2012). Returning to chapters 1 and 2, we can think of how some apparatuses function atmospherically and act on and through atmospheres (particularly in relation to modalities of power based on environmentalities). In the spaces of torture that connect the Cold War and the War on Terror, atmospheres are weaponised as modalities of sovereign and environmental power meet. For example, in Chapter 3's discussion of 'debility dependency dread', the 'Alice in Wonderland' interrogation technique attempted to create an affective atmosphere of confusion that disorientated prisoners. The interrogation manual stating that: 'In this strange atmosphere the subject finds that the pattern of speech and thought which he has learned to consider normal have been replaced by an eerie meaninglessness' (KUBARK 1963: 76).

In the final scene we move to a different site: an area outside the control room where 'tactical' response is being rehearsed at the site of a fire. Ten minutes in something happens and an atmosphere emanates, lives for a while and dies. And in this process we get a sense of something Dufrenne and Böhme teach us: that every atmosphere, even if it might resemble others, even if we might use a conventional vocabulary to name and know it, and even if it is made through a series of conventional arrangements, is singular: unrepeatable. It comes and goes:

> "He's fucking overdoing it now", the police officer says to me as we stand watching the scene in front of us. Exasperated, his colleague who is also observing adds: "it's pointless, this". We've changed locations to where the "tactical" dimensions of response are being played. In front of us a "member of the public" is refusing to follow the instructions given by members of the fire and rescue service. Instead, he steals anything red. This is unfortunate as most the fire and rescue services equipment is red. Now going on for five minutes, two fire officers move him from the scene of the explosion. Initially, the people observing laughed at his antics. Now they frown, tut, exchange comments, and mutter (Author's Fieldnotes from Exercise Observation, 5 June 2010).

6.6 Affective Conditions

Atmospheres may emanate from bodies but they are not reducible to them. This is Dufrenne and Böhme's point. The singular affective qualities that are atmospheres – homely, serene, erotic, urgent, exciting, and so on – exceed the ensembles from which they emanate. They are irreducible. Atmospheres are a kind of indeterminate affective excess through which intensive space-times are created and come to envelop specific bodies; sites, objects, people, and so on, all may be atmospheric or may feel and be moved by atmospheres. The vague sense of atmosphere as a 'more' seems an appropriate place to finish this chapter and the discussion of collective affects as environments that may condition how life is lived and felt. For me, the concept of atmosphere is good to think with because it holds a series of opposites – presence and absence, cause and effect, subject and object – in a relation of tension. We feel this tension if we juxtapose Marx's materialist imagination with a phenomenology attentive to singular affective qualities. By invoking a material imagination based on the movement and lightness of air, we learn from the former about the turbulence of atmospheres and their indeterminate quality. From the work of Böhme and Dufrenne, we learn that atmospheres are singular affective qualities that emanate from particular bodies – 'total effects' that envelop but are irreducible to ensembles.

All this means that inseparable from the concept of atmosphere is a question: how is it possible to sense collective affects that although not reducible to the bodies that they envelop and effect are always being differentially expressed and qualified (or, in Dufrenne's terms, 'gain being' from a 'plurality of interpretations')? The description of complex forms of causality in the final section of the chapter is an attempt to sense the variety of ways in which things and people can affect and be affected through atmospheres, as well as holding onto the strange reality of atmospheres. As such, the chapter should be read alongside the speculations on structures of feeling offered in the previous chapter. Structures of feeling link different sites, occurring across them, and create something like a predisposition to self, others and the world. Affective atmospheres envelop and emanate from particular ensembles that are gathered together for different durations around particular bodies. Turning back to the discussion in Chapter 4 of how a body's 'force of existing is formed in encounters, we can understand both affective atmospheres and structures of feeling as processes of mediation that mix the formed and formless, emergent and finished, structural and ephemeral.

When taken together, structures of feeling and affective atmospheres are two partially connected ways of thinking through the affective conditions that mediate how encounters take place and how apparatuses form and operate. My aim has been to move beyond any assumption that definable cultures, or periods of time, have characteristic moods without losing the sense that affective life extends beyond the boundaries of individual subjects and the coordinates of punctual encounters. Neither can collective affects be reduced to the incessant movement of pre-personal intensities endlessly circulating. Condition may seem an odd concern to resurrect, linked as it is in the social science to a search for some form of transcendent cause that stands apart from practical action and determines socio-spatial life (think of the determining role sometimes afforded to 'economic conditions' in the last instance). Whilst noting that risk, I use the term condition because of the complex sense of mediation that the term offers. A condition does not determine. Rather, structures of feeling and affective atmospheres condition in the sense of being presences that affect something outside of themselves and becoming mediums through which ordinary affective life is lived and organised. Consequently, affective conditions are constantly being expressed in feelings and qualified in emotions, existing in a tensed space-time between the finished and unfinished, the ephemeral and the structural, the determinate and indeterminate. This means that even as they mediate encounters and apparatuses, structures of feeling and affective atmospheres may be reworked during encounters and shaped in and through the operations of apparatuses (including but not limited to signifying apparatuses that organise affective investments and relations). Affective conditions are never self-evidently present, as we may sometimes assume other conditions are. They are always in the midst of encounters, emerging and changing as they mediate life, shaping how the world is disclosed, related to and felt, and becoming inseparable from affects and emotions.

Chapter 7
Mediating Affective Life

> all active relations between different kinds of beings and consciousness are inevitably mediated, and this process is not a separable agency – a 'medium' – but intrinsic to the properties of the related kinds. ... Thus mediation is a positive process in social reality, rather than a process added to it by way of projection, disguise, or interpretation
>
> (Williams 1977: 98–9)

7.1 An Analytics of Affect

At the heart of this book is Eve Kosofsky Sedgwick's (2003) insight that affects *may* attach to more-or-less anything: institutions, people, things, and so on. The first implication of Sedgwick's axiom is that there can never be a securely bounded realm of affectual/emotional geography happily and safely separate from other well-recognised areas of analysis. The other implication is that work on affect needs to be capable of disclosing how affective life differs as attachments are made and remade. For that reason, I have explored how multiple processes of mediation order affective life, even whilst I have stressed that affect is irreducible to those processes. Mediation is understood in the double sense that Williams outlines in this chapter's epigraph. First, affective life is always-already 'in the midst of' relations and processes and inseparable from those relations and processes. Second, processes of mediation are active in the sense that affective life differs as it is mediated. Differences are not a secondary reaction or response to the specific forms, types and channels of organisation that produce patterns of affective life. Instead, differences in affects, feelings, emotions, structures of feeling and atmospheres emerge through processes of ordering. This is why my focus has been horizontal to how bodies come together and apart rather than to the non-conscious dynamics of the neurological or biophysical or psychoanalytical body. For it is in how collectives are gathered – the ordering effects of apparatuses, encounters or conditions – that affects, feelings and emotions differ. Affective atmospheres or structures of feeling may be one part of the conditions through which affective life is mediated. Other processes of mediation might include how people have learnt to encounter one another, how affective life is known, rendered actionable and intervened in, or how affects condense around specific, socially differentiated, bodies. In short, my concern has been with *some* of the ways in which a body's 'force of existing', its capacity to affect and be affected, is mediated.

I have explored the mediation of affect at a time of flourishing theoretical interest in affect and emotion as various disciplines learn to attend to the dynamics of life and living. In the midst of all this I have assembled one version, to borrow

Despret's (2004) term, of affect; or, better, one way to orientate to and attune to affective life. It is only one version though. It coexists with others. It is indebted to work that first emphasised that emotions matter, in particular feminist work on the politics of emotional life. It is also incomplete. I have deliberately not attempted to review every theory of affect or emotion, nor cover every possible issue. Like any version, the account of affective life has its own geographically and historically contingent conditions of formation in relation to a diverse set of topics, problems and concerns. The examples I have included in the book were designed to reflect the breadth of those specific conditions as well as exemplify that affects can be found throughout life; sites of violence, everyday events of hope, neo-liberalising apparatuses and the politics of emergency. As a version, then, the account of affective life has the status of a proposition; one claim amongst others, to be encountered…

The book is also a response to a set of problematics that strike me as key to understanding how affective life happens. My aim has been to assemble a vocabulary that has a chance of understanding some of the ways in which affective life is patterned as it is (re)ordered. This has meant trying to hold onto the way in which affective life is simultaneously mediated in the sense that it reflects and expresses relations, and mediating in that affects, structures of feelings and atmospheres are always imbricated with other processes. Rather than the theory of affect, what I have offered is better described as an analytics of affect, despite the seeming coldness of the term 'analytic' – after Foucault's (1978: 82) comments on the task of his work on power.[1] For what is produced is a vocabulary (apparatus, version, affect, structures of feeling, atmosphere) that treats affects as simultaneously: an *object-target* of apparatuses; a *bodily capacity* emergent from encounters and a *collective condition* that mediates how life is lived and felt. Underpinning the orientation to affect an object-target, bodily capacity and collective condition has been an attention to how ordering happens in a diverse world riven by human and non-human differences. Whether thinking about apparatuses as the integration of a differential field, encounters as transpersonal or the emanation of atmospheres from an ensemble, my emphasis has been on ordering as an ongoing precarious achievement through which affect is continually (re)contextualised. I have found inspiration for such an approach in a variety of non-representational theories. For what non-representational theories share is a commitment to following how ordering happens immanently rather than through some form of transcendent structure. Too often in work on the politics of affect, some form of context is lazily invoked as a kind of explanatory backdrop to analysis. What is common to the theoretical resources I have assembled in this book is that they avoid this shortcut by taking as their starting point the problem of arrangements and wondering away

1 The quote is from Foucault (1978: 82) on power. As is now well known, Foucault explains that his task is to work: 'toward an "analytics" of power: that is, toward a definition of the specific domain formed by relations of power, and toward a determination of the instruments that will make possible its analysis'.

at how affective life is mediated. To conclude, I turn directly to the question of mediation in order to rework my four starting points outlined in Chapter 1 and suggest what an analytics of affect might do.

7.2 Mediating Affective Life

The affective turn in the social sciences and humanities has been inseparable from a renewed concern with the multifaceted concepts of life and living. Whilst varied, the emphasis has been in part, but not exclusively, on how beginning from an ontological version of affect discloses a processual world of movement.[2] The self-possessive individual is dissolved into a field of events and relations and replaced at the heart of analysis by a concept of life that centres on change. Affective life is then treated as exuberant, mutable and ever generative of newness. The subject is a temporary extraction from life whose formation is always provisional. Countering the reduction of life in certain accounts of biopower, theorists have argued that this affirmative vision of life requires a practice of research that is vital and creative. This approach has generated many insights, but my starting point has been different. I have tried to avoid starting from 'affect itself' or affect 'as such'. Which is another way of trying not to give some kind of positive value to life's generativity. Affect does not provide a window onto, or an index of, life's force and energy. Chapter 4– The Imbrication of Affect – was so titled to make it clear that affective life is always-already involved with what might seem non-affective. In short, and to put it simply, affective life is mediated. This is something well known from social constructionist accounts of emotion, even if mediation has been reduced to one particular form: signifying apparatuses (Grossberg 2010). In discussing three translations of affect – as object-target, bodily capacity and collective condition – I have attempted to specify the operation of qualitatively different processes of mediation whilst holding onto the irreducibility of affect. These processes include signifying apparatuses, but are not reducible to them. Let us summarise the processes of mediation specific to affect as object-target, bodily capacity and collective condition in turn, recognising that there will also be other processes of mediation that work to articulate and contextualise affect (Grossberg 1997). As Grossberg (2010: 189) stresses, one critical task for analysis is to multiply the 'modalities, practices and agencies of mediation' beyond signification (which is not the same as ignoring how life is mediated through signification). Doing so gives us cause to pause before affirming either

2 Brian Massumi (2002a: 8) has been the most explicit proponent of an equivalence between affect and movement. He stresses: 'Position no longer comes first, with movement a problematic second. It is secondary to movement and derived from it. It is retro movement, movement residue. The problem is no longer to explain how there can be change given positioning. The problem is to explain the wonder that there can be stasis given the primacy of process'.

an ontology that presumes affect's inevitable escape over and above the various ways in which life is organised, or valuing emotion as an authentic expression of a coherent subject that can be or should be recovered.

Understanding how affect is known, rendered actionable and intervened in requires that we attend to the composition and operation of apparatuses. This is the first way in which affective life is mediated: through ways of knowing closely linked to interventions that come to 'inform' (Barry 2006) affective life. Distinguishing my approach from claims that 'affect itself' is only now an object-target of new forms of power, my emphasis in chapters 2 and 3 was on the life of apparatuses. Integrating ways of knowing affective life with practices of intervention, distinct versions of affective life are produced as affects are propertied and bounded as particular kinds of things. Apparatuses involve diverse forms of action on and through affect, rather than simply the manipulation of affective subjects shorn of the capacity to reason and addressed at the level of a technically mediated biomediated body. This means that the specific form of mediation will vary depending on the apparatus. Disciplinary apparatuses, for example, involve the entraining of optimal sequences of action, whereas environmentalities involve the setting up of an environment within a set of thresholds. Starting from apparatuses as the 'integration of a differential field' allows us to avoid an epochal analysis of the relation between power in the early twenty-first century and affective life. Instead, the critical task is to trace the specific connections between how life is known, rendered actionable and intervened in and modalities of power. As we saw in the analysis of morale or 'debility, dependency, dread', those modalities of power are multiple, partially connected, and are riven by their own volatility. Versions of affect change, apparatuses change and both have an affective life. It is a mistake, then, to counterpoise efforts to act on and through affective life, whether through manipulation or some other mode of action, to the dynamism of affective life. Apparatuses are dynamic processes whereby a multiplicity is integrated, they have their own volatility and mutability even if they may be animated by an urgent need, and are elaborated in relation to a not-yet determined future.

Affective life is also mediated in and through encounters. Emphasising how a body's 'force of existing' emerges, is sustained and changes through encounters draws attention to how affective life happens in relation. Rejecting an analytic divide between emotion, feeling and affect, we can think of something like hope as an emergent affective property of encounters. Both personal and impersonal, affects are mediated by everything from an encounter with the memory of a lost loved one, through to an event such as unemployment or a materiality such as music. Anchoring Chapter 4 in a case of hope was designed to exemplify a move from the analysis of 'affect itself' to the imbrication and irreducibility of affect. Rather than focus on preconscious bodily physiology or psychology, this involves an attunement and orientation to what a body can do as mediated through relations, some of which will extend beyond the encounter, folding it into other times and spaces.

Encounters are not simply punctual: bodies enter any encounter with an existing 'force of existing' and encounters always relate to other times and spaces. Furthermore, a body's 'force of existing' is not simply bodily and autonomic. It may be constituted through, for example, habitual practices of reflection where the act of reflecting becomes an intentional, or not, means of amplifying or dampening feeling. Or a body's 'capacity to affect and be affected' may be changed through, say, participation in an intentional practice of remembering in which a forgotten past comes to be relived, if only momentarily before fading into an atmosphere. Mediation may also occur through the multiple speech acts and representational acts that participate in encounters and come to have an affective force. In short, I have wanted to cut the equation between affect and the momentary encounter and orientate attention to some of the different ways in which bodily capacities are organised. Any opening up to life's potentiality, the 'more to life' to paraphrase Massumi (2002a), must therefore be understood as part of a pattern of encounters, as well as the geo-historicity of the body. Encounters must also be understood in relation to the life of apparatuses, which may shape what participates in an encounter and how encounters happen. This does not mean that affective life is simply determined by power. That would be to raise power to the status of quasi-mythical substance. To paraphrase Foucault's (1978) wonderful comments on the relation between life and power, affects may be instruments or effects of apparatuses, but also they may be impediments or irritants. Ordinary fears, hopes or other affects may also be the starting point for alternative strategies, a site of reproduction, or may be outside of the sphere of a particular modality of power's operation and organisation.

It would be easy to juxtapose an emphasis on affect as an object-target with an emphasis on affect as a bodily capacity emergent from encounters: where the former highlights all the ways in which life is reduced, and the latter orientates us to life's excesses. In emphasising the life of apparatuses and the patterning of encounters I have tried to think outside of this one-dimensional relation. Introducing a third translation of affect further reworks that relation: affect is a collective condition that mediates encounters and apparatuses and is, in turn, mediated. If apparatuses are one way in which encounters are mediated, then collective affects can condition and be reworked through both encounters and apparatuses. My wager is that the concepts of structures of feeling and affective atmospheres allow us some purchase on the reality of collective affects – whether by showing up certain things as being of significance or by giving activities a specific affective quality. Differently expressed and qualified in bodily capacities, and so always being translated in encounters, I have nevertheless attempted to treat structures of feeling and affective atmospheres as real collective phenomenon that mediate environmentally. By which I mean that we are in collective affects that shift between absence and presence; being enveloped by the fleeting mood of a room, feeling the anger of a crowd, enraged by the tone of a book on affect. Collective affects mediate by shaping how self, others and the world can be and are related to and made present. Due emphasis must be placed on how affective atmospheres or structures of feelings are qualified in bodily capacities and articulated with signifying apparatuses. By which I mean that we need to keep hold of the double status of collective affects: they are at once mediators in the sense that they shape affective life, and mediated in the sense that they are themselves shaped by particular relations, events and things. We might consider, for example, how collective fears can be mediated by signifying apparatuses which, as Grossberg (2010) shows, often work through binary logics, negate otherness and distribute bodily capacities. But I have also tried to highlight the irreducibility of affective conditions: affective conditions are not secondary phenomena explainable by reference to some other social force or factor.

Mediating processes happen together and are entangled: apparatuses relate to other apparatuses, affective atmospheres merge and blur with one another, encounters are articulated with structures of feeling that disclose life, the things that populate encounters are enveloped by particular atmospheres, and so on and so on. The focus for analysis could, then, be complex spatial-temporal formations that involve specific combinations of different processes and forms of affective life. Undoubtedly, there are other processes of mediation that I have only touched upon, hinted at or ignored (most notably how language and communication mediate affective expression (Brown and Stenner 2009)). My aim in this book has not been to offer a comprehensive theory of affect that would presume to be the last word. I have tried to remain open to further elaboration and development. This has involved moving beyond the now familiar starting point that affects and/or emotions are relational phenomena and/or relationally constituted, whilst holding onto the key insight of that work: that affective life happens in the middle of relations to the self, others or the world and relations take many forms. Likewise, I have wanted to learn from the attention to life as impersonal force, without giving life's productivity a positive value or finding in life a guarantee that things might be different and better or necessarily oppositional to modalities of power. Apparatuses, encounters and conditions mediate affective life – in the sense that affective life is constantly being made and remade through an array of processes that produce differences in what is and can be felt. And those forms of mediation should be thought of as active processes in Williams's (1977) sense, albeit in different ways. They do not only reduce life's exuberance, nor simply capture and close affect.

We are now in a position to return to the propositions I set out in Chapter 1 in light of these comments on multiple mediations. The propositions become questions to guide an analytics of affect that centres on understanding mediation: the array of processes that shape affective life and result in temporary orderings of affect understood as object-target, bodily capacity and collective condition.

1. If there is no such thing as 'affect itself', then affects are always being contextualised and articulated with … more or less anything. One key task for analysis is to understand the operation, efficacy and coexistence of the varied processes of mediation through which affective life is ordered.

For example, what is the relation between a body's 'force of existing', a signifying apparatus and the things that make up an encounter?

2. If affective life is collective because it is always-already mediated, then we need to understand how affective life is part of specific patterns of relations and events. For example, how are the apparatuses through which affective life is known organised around a strategic function and urgent need? How do atmospheres emanate as the 'total effect' of ensembles? How are collective affects simultaneously structured and diffuse? Finally, how do all kinds of bodies come together in the encounters that reproduce, express and enact bodily capacities? Apparatuses, encounters, structures and ensembles are all different ways in which affective life is organised.
3. If affects, feelings, atmospheres, and so on, are irreducible to other non-affective social or natural processes, then analysis might focus on the different and specific ways in which they act and make a difference. How do collective affects mediate – whether as an atmosphere that envelops or a structure of feeling that disposes? And how does someone's 'force of existing' become part of encounters and shape how encounters happen and are felt?
4. If affect is not the non-representational object per se, then language and communication become one form of 'positive process in social reality' (Williams 1977: 98) amongst others, and representations of affective life may have performative effects. Looping back to chapters 2 and 3, an analytics of affect might ask how affective life is always-already informed by 'versions' of emotion, affect, feeling, passion, and so on (as well as considering representations as active elements in encounters, apparatuses and conditions). A theory of affect that acknowledges the multiplicity of versions of affect that exist must also acknowledge its own contingency.

The questions orientate an analytics of affect to the complexity of how affective life happens, and to the emergence of differences as affective life is continually mediated. An analytics of affect affirms that atmospheres, structures of feeling, and so on, may suffuse all of life, are irreducible, and may attach to anything – sometimes in surprising ways, often not.

Bibliography

Adey, P. (2010) *Aerial Life: Spaces, Mobilities, Affects.* London: Wiley-Blackwell.

Agamben. G. (2009) What is an Apparatus? In Agamben, G. *What is an Apparatus? And Other Essays.* Translated by Kishik, D. and Pedatella, S. Stanford, CA: Stanford University Press, 1–24.

Ahmed, S. (2004) *The Cultural Politics of Emotion.* London: Routledge.

Ahmed, S. (2007–2008) Multiculturalism and the promise of happiness. *New Formations* 63, Winter, 121–37.

Ahmed, S. (2010) *The Promise of Happiness.* London: Duke University Press.

Aitken, S. (2009) *The Awkard Spaces of Fathering.* Farnham: Ashgate.

Amoore, L. (2011) Data derivatives: on the emergence of a security risk calculus for our times. *Theory, Culture & Society* 28, 24–43.

Anderson, B. (2004a) Recorded music and practices of remembering. *Social and Cultural Geography* 5, 1–19.

Anderson, B. (2004b) Time-stilled space-slowed: how boredom matters. *Geoforum* 35, 739–54.

Anderson, B. (2005) Practices of judgement and domestic geographies of affect. *Social and Cultural Geography* 6(6), 645–60.

Anderson, B. (2006) Becoming and being hopeful: towards a theory of affect. *Environment and Planning D: Society and Space* 24, 733–52.

Anderson, B. (2007) Hope for nanotechnology: anticipatory knowledge and the governance of affect. *Area* 39(2), 156–65.

Anderson, B. (2012) Affect and biopower: Towards a theory of affect. *Transactions of the Institute of British Geographers* 37(1), 28–43.

Anderson, B. and Adey, P. (2012) Governing events and life: 'emergency' in UK civil contingencies. *Political Geography* 31(1), 24–33.

Anderson, B. and Harrison, P. (2010) The Promise of Non-representational Theories. In Anderson, B. and Harrison, P. (eds) *Taking-Place: Non-representational Theories and Geography.* Farnham: Ashgate, 1–36.

Anderson, B. and Wylie, J. (2009) On geography and materiality. *Environment and Planning A* 41(2), 318–35.

Anderson, K. and Smith, S. (2001) Emotional geographies. *Transactions of the Institute of British Geographers* 26, 7–10.

Angell, J. (1941) Radio and national morale. *The American Journal of Sociology* 47(3), 352–59.

Aradau, C. and Van Munster, R. (2011) *The Politics of Catastrophe: Genealogies of the Unknown.* Abingdon: Routledge.

Arendt, H. (1958) *The Human Condition.* Chicago: University of Chicago Press.

Ash, J. (2010) Architectures of affect: anticipating and manipulating the event in practices of videogame design and testing. *Environment and Planning D: Society and Space* 28(4), 653–71.

Ash, J. (2012) Attention, videogames and the retentional economies of affective amplification. *Theory, Culture & Society* 29(6), 3–26.

Barad, K. (2003) Posthumanist performativity: toward an understanding of how matter comes to matter. *Signs: Journal of Women in Culture and Society.* 28(3), 801–31.

Barad, K. (2007) *Meeting the Universe Halfway.* Durham, NC and London: Duke University Press.

Barnett, C. (2008) Political affects in public space: normative blind-spots in non-representational ontologies. *Transactions of the Institute of British Geographers* 33(2), 186–200.

Barry, A. (2006) Pharmaceutical matters: the invention of informed materials. *Theory, Culture & Society* 22(1), 51–69.

Bateson, G. (1942) Morale and the National Character. In: Watson, G. (ed.) *Civilian Morale.* Second Yearbook of the Society for the Psychological Study of Social Issues. New York: Reynal and Hitchcock, 71–91.

Bauman, Z. (2006) *Liquid Fears.* London: Polity Press.

Benjamin, W. (1969) Thesis on the Philosophy of History. In: Benjamin, W. *Illuminations.* Translated by Zorn, H. London: Pimlico, 245–55.

Bennett, J. (2001) *The Enchantment of Modern Life.* Princeton and London: Princeton University Press.

Bennett, J. (2005) The agency of assemblages and the North American blackout. *Public Culture* 17(3), 445–65.

Bennett, J. (2010) *Vibrant Matter: A Political Ecology of Things.* Durham, NC and London: Duke University Press.

Berlant, L. (2004) Introduction. In: Berlant, L. (ed.) *Compassion: The Culture and Politics of an Emotion.* New York: Routledge, 1–13.

Berlant, L. (2011) *Cruel Optimism.* Durham NC and London: Duke University Press.

Bissell, D. (2008) Comfortable bodies: sedentary affects. *Environment and Planning A* 40(7), 1697–1712.

Bissell, D. (2010) Passenger mobilities: affective atmospheres and the sociality of public transport. *Environment and Planning D: Society and Space* 28(2), 270–89.

Blackman, L. (2012) *Immaterial Bodies: Affect, Embodiment, Mediation.* London: Sage.

Blackman, L. and Venn, C. (eds) (2010) Special issue: Affect. *Body and Society* 16, 1.

Bloch, E. (1986) *The Principle of Hope (Volumes 1–3).* Translated by Plaice, N., Plaice, S. and Knight, P. Oxford: Blackwell.

Bloch, E. (1998) Can Hope be Disappointed? In: *Literary Essays.* Translated by Joron, A. Stanford, CA: Stanford University Press, 339–45.

Böhme, G. (1993) Atmosphere as the fundamental concept of a new aesthetics. *Thesis Eleven* 36, 113–26.

Böhme, G. (1995) *Atmosphäre. Essays zur neuen Ästhetik*. Frankfurt am Main: Suhrkamp.

Böhme, G. (2006) Atmosphere as the Subject Matter of Architecture. In: Ursprung, P. (ed.) *Herzog and Meuron: Natural History.* London: Lars Müller Publishers, 398–407.

Boler, M. (1999) *Feeling Power: Emotions and Education.* London: Routledge.

Boltanski, L. and Chiapello, E. (2005) *The New Spirit of Capitalism.* Translated by Elliott, G. London: Verso.

Bondi, L. (2005) Making connections and thinking through emotions: between geography and psychotherapy. *Transactions of the Institute of British Geographers* 30(4), 433–48.

Bondi, L., Davidson, J. and Smith, M. (2005) Introduction: Geography's Emotional Turn. In: Davidson, J., Bondi, L. and Smith, M. (eds) *Emotional Geographies.* Farnham: Ashgate, 1–16.

Borch, C. (2008) Foam architecture: managing co-isolated associations. *Economy and Society* 37(4), 548–71.

Borchelt, G. (2005) *Break Them Down: Systematic Use of Psychological Torture by US Forces.* Cambridge, MA and Washington: Physicians for Human Rights.

Boulnois, O. (2006) Object. *Radical Philosophy* 139, September/October, 123–33.

Brennan, T. (2004) *The Transmission of Affect.* Ithaca and London: Cornell University Press.

Brown, S.D. and Stenner, P. (2009) Psychology Without Foundations. In: *History, Philosophy and Psychosocial Theory.* London: Sage.

Burke, J. (1999) *An Intimate History of Killing: Face-to-Face Killing in Twentieth-Century Warfare.* London: Granta.

Butler, J. (2004) *Precarious Life: The Powers of Mourning and Violence.* New York: Verso.

Cadman, L. (2009) Non-representational Theory/Non-representational Geographies. In: Kitchin, R. and Thrift, N. (eds) *International Encyclopaedia of Human Geography*. Amsterdam: Elsevier, 456–63.

Caputo, J. (2006) *The Weakness of God: A Theology of the Event.* Indiana: Indiana University Press.

Carter, S. and McCormack, D. (2006) Film, geopolitics and the affective logics of intervention. *Political Geography* 25(2), 225–45.

Casarino, C. (2008) Time Matters: Marx, Negri, Agamben, and the Corporeal. In: Casarino, C. and Negri, A. *In Praise of the Common: A Conversation on Philosophy and Politics.* Minneapolis: University of Minnesota Press, 219–46.

Castel, R. (1991) From Dangerousness to Risk. In: Burchell, G., Gordon, C. and Miller, P. (eds) *The Foucault Effect.* Chicago: University of Chicago Press, 281–98.

Chickering, R., Förster, S. and Greiner, B. (eds) (2004) *A World at Total War: Global Conflict and the Politics of Destruction, 1937–1945.* Cambridge: Cambridge University Press.

Chow, R. (1999) The Politics of Admittance. Female Sexual Agency, Miscegenation, and the Formation of Community in Frantz Fanon. In: Alessandrini, A. (ed.) *Frantz Fanon: Critical Perspectives.* London: Routledge, 34–56.

Clough, P. (2004) Future matters: technoscience, global politics and cultural criticism. *Social Text* 22(3), 1–23.

Clough, P. (2007) Introduction. In: Clough, P. with Halley, J. (eds) *The Affective Turn. Theorizing the Social.* Durham, NC and London: Duke University Press, 1–33.

Clough, P. (2008) The affective turn: political economy, biomedia and bodies. *Theory, Culture & Society* 25(1), 1–22.

Clough, P. with Halley, J. (eds) (2007) *The Affective Turn. Theorizing the Social.* Durham NC and London: Duke University Press.

Collier, S. (2009) Topologies of power: Foucault's study of political government beyond 'governmentality'. *Theory, Culture & Society* 26, 78–108.

Colls, R. (2013) Feminism, bodily differences and non-representational geographies. *Transactions of the Institute of British Geographers* 37, 430–45.

Connolly, W.E. (2002) *Neuropolitics: Thinking, Culture, Speed.* Minneapolis: University of Minnesota Press.

Connolly, W.E. (2004) Method, Problem, Faith. In: Shapiro, I., Smith, R. and Masoud, T. (eds) *Problems and Methods in the Study of Politics.* Cambridge: Cambridge University Press, 332–49.

Connolly, W.E. (2005) *Pluralism.* Durham, NC and London: Duke University Press.

Connolly, W.E. (2011) *A World of Becoming.* Durham, NC and London: Duke University Press.

Conradson, D. (2010) The Orchestration of Feeling: Stillness, Spirituality and Places of Retreat. In: Bissell, D. and Fuller, G. (eds) *Stillness in a Mobile World.* London: Routledge, 71–86.

Cusick, S. (2006) 'Music as torture/music as weapon'. Available at: http://www.sibetrans.com/trans/a152/music-as-torture-music-as-weapon (last accessed 20 June 2012).

Cusick, S. (2008) 'You are in a place that is out of the world … ': music in the detention camps of the 'Global War on Terror'. *Journal of the Society for American* Music 2(1), 1–26.

Danner, M. (2004) *Torture and Truth: America, Abu Ghraib, and the War on Terror.* London: Granta.

Davidson, J. (2003) 'Putting on a face': Sartre, Goffman, and agoraphobic anxiety in social space. *Environment and Planning D: Society and Space* 21, 107–22.

Davis, M. (1999) *Ecology of Fear: Los Angeles and the Imagination of Disaster.* Los Angeles: Vintage Books.

De Goede, M. and Randalls, S. (2009) Precaution, preemption: arts and technologies of the actionable future. *Environment and Planning D: Society and Space* 27(5), 859–78.

De Landa, M. (1991) *War in the Age of Intelligent Machines.* London: Zone Books.

De Landa, M. (2005) *A New Philosophy of Society. Assemblage Theory and Social Complexity.* London and New York: Continuum.

Deleuze, G. (1978) 'Lecture of 24.01.1978' (translator unknown). Available at www.webdeleuze.comTXT/ENG/240178 (last accessed 23 March 2002).

Deleuze, G. (1986) *Cinema 1: The Movement Image.* Translated by Tomlinson, H. and Habberjam, B. London: Athlone Press.

Deleuze, G. (1988a) *Spinoza: Practical Philosophy.* Translated by Hurley, R. San Francisco: City Lights Books.

Deleuze, G. (1988b) *Foucault.* Translated by Hand, S. London and New York: Continuum.

Deleuze, G. (1991) *Bergsonism.* Translated by Tomlinson, H. and Habberjam, B. New York: Zone Books.

Deleuze, G. (2001) *Pure Immanence: Essays on a Life.* Translated by Boyman, A. New York: Zone Books.

Deleuze, G. (2008) What is a 'Dispositif'? In: Lapoujade, D. (ed.) *Two Regimes of Madness: Texts and Interviews, 1975–1995.* New York: Semiotext(e), 338–48.

Deleuze, G. and Guattari, F. (1987) *A Thousand Plateaus.* Translated by Massumi, B. London: Continuum.

Deleuze, G. and Guattari, F. (1994) *What is Philosophy?* Translated by Tomlinson, H. and Burchill, G. London and New York: Verso.

Deleuze, G. and Parnet, C. (2003) *Dialogues II.* Translated by Tomlinson, H. and Habberjam, B. London: Continuum.

Despret, V. (2004) *Our Emotional Makeup: Ethnopsychology and Selfhood.* Translated by De Jager, M. New York: Other Press.

Dewsbury, J-D. (2000) Performativity and the event: enacting a philosophy of difference. *Environment and Planning D: Society and Space* 18(4), 473–97.

Dewsbury, J-D. (2007) Unthinking subjects: Alain Badiou and the event of thought in thinking politics. *Transactions of the Institute of British Geographers* 32(4), 443–59.

Dewsbury, J-D. (2009) Performative, Non-representation and affect-based research: seven injunctions. In: Delyser, D., Aitken, S., Crang, M., Herbert, S. and McDowell, L. (eds) *The SAGE Handbook of Qualitative Geography.* London and New York: Sage, 321–34.

Dewsbury, J-D. (2010) Language and the event: the unthought of appearing worlds. In: Anderson, B. and Harrison, P. (eds) *Taking Place: Non-Representational Theories and Geography.* Farnham: Ashgate, 147–60.

Dewsbury, J-D., Harrison, P., Rose, M. and Wylie, J. (2002) Enacting geographies. *Geoforum* 33(4), 437–40.

Diken, B. (2008) Climates of nihilism. *Third Text* 22(6), 719–32.

Dillon, M. and Reid, J. (2009) *The Liberal Way of War: Killing to Make Life Live.* London: Routledge.

Diprose, R. (2001) *Corporeal Generosity: On Giving with Nietzsche, Merleau-Ponty, and Levinas.* New York: SUNY University Press.

Dixon, T. (2003) *From Passions to Emotions: The Creation of a Secular Psychological Category.* Cambridge: Cambridge University Press.

Doel, M. (2004) Poststructuralist Geographies: The Essential Selection. In: Cloke, P., Crang, P. and Goodwin, M. (eds) *Envisioning Human Geography.* London: Arnold, 146–71.

Doel, M. (2010) Representation and Difference. In: Anderson, B. and Harrison, P. (eds) *Taking-Place: Non-Representational Theories and Geography.* Farnham: Ashgate, 117–30.

Dolan, F. (1994) *Allegories of America: Narratives, Metaphysics, Politics.* Ithaca: Cornell University Press.

Donzelot, J. (1991) Pleasure in Work. In: Burchell, G., Gordon, C. and Miller, P. (eds) *The Foucault Effect: Studies in Governmentality.* Chicago: University of Chicago Press, 251–80.

Douhet, G. (1972) *Command of the Air.* New York: Arno Press.

Dufrenne, M. (1976) *The Phenomenology of Aesthetic Experience.* Translated by Casey, E., Anderson, A., Domingo, W. and Jacobson, L. Evanston: Northwestern University Press.

Durant, H. (1941) Morale and its measurement. *The American Journal of Sociology* 47(3), 406–14.

Dwyer, C., Shah, B. and Sanghera, G. (2008) From cricket lover to terror suspect: challenging representations of young British Muslim men. *Gender, Place and Culture* 15(2), 117–36.

Eagleton, T. (1980) *Criticism and Ideology: A Study in Marxist Literary Theory.* London: Verso.

Ehrenreich, B. (1997) *Blood Rites: Origins and History of the Passions of War*. New York: Metropolitan Books.

Elden, S. (2001) *Mapping the Present: Heidegger, Foucault, and the Project of a Spatial History*. London: Continuum.

Esposito, R. (2008) *Bios: Biopolitics and Philosophy*. Minneapolis: University of Minnesota Press.

Estorick, E. (1941) Morale in contemporary England. *The American Journal of Sociology* 47(3), 462–71.

Evans, B. (2010) Anticipating fatness: childhood, affect, and the pre-emptive 'war on obesity'. *Transactions of the Institute of British Geographers* 35(1), 21–38.

Fanon, F. (1986) *Black Skin, White Masks.* Translated by Markmann, C. London: Pluto Press.

Farago, L. (1941) (ed.) *German Psychological Warfare: Survey and Bibliography.* New York: Committee for National Morale.

Farber, I., Harlow, H. and West, L. (1957) Brainwashing, conditioning, and DDD (debility, dependency, and dread). *Sociometry* 20(4), 271–85.

Farish, M. (2010) *The Contours of America's Cold War*. Minneapolis: University of Minnesota Press.

Feher, M. (2009) Self-appreciation; or, the aspirations of human capital. *Public Culture* 21(1), 21–41.

Fisher, M. (2009) *Capitalist Realism. Is There No Alternative?* Winchester and Washington: Q Books.
Flatley, J. (2008) *Affective Mapping: Melancholia and the Politics of Modernism.* Cambridge, MA: Harvard University Press.
Fortunati, L. (1995) *The Arcane of Reproduction.* Translated by Creek, H. and Fleming, J. New York: Autonomedia.
Fortunati, L. (2007) Immaterial labour and its machinization. *Ephemera: Theory and Politics in Organisation* 7(1), 139–57.
Foucault, M. (1977) *Discipline and Punish.* Translated by Sheridan, A. London: Penguin Books.
Foucault, M. (1978) *The History of Sexuality: Volume One.* Translated by Hurley, R. London: Penguin Books.
Foucault, M. (1980) The Confession of the Flesh. In: Gordon, C. (ed.) *Power/ Knowledge: Selected Interviews and Other Writings.* London: The Harvester Press, 194–228.
Foucault, M. (1988) Practicing Criticism. In: Kritzman, L. (ed.) *Politics, Philosophy, Culture: Interviews and Other Writings, 1977–1984.* Translated by Sheridan, A. and others. New York: Routledge, 153–56.
Foucault, M. (1994a) The Political Technology of Individuals. In: Faubion, J. (ed.) *Power: Essential works of Foucault 1954–1984, Volume Three.* Translated by Hurley, R. and others. London: Penguin, 403–17.
Foucault, M. (1994b) 'Omnes et Singulatim': Towards a Critique of Political Reason. In: Faubion, J. (ed.) *Power: Essential works of Foucault 1954–1984, Volume Three.* Translated by Hurley, R. and others. London: Penguin, 298–325.
Foucault, M. (1994c) The Politics of Health in the Eighteenth Century. In: Faubion, J. (ed.) *Power: Essential Works of Foucault 1954–1984, Volume Three.* Translated by Hurley, R. and others. London: Penguin, 90–105.
Foucault, M. (1996) What is Critique? In: Schmidt, J. (ed.) *What is Enlightenment? Eighteenth Century Answers and Twentieth Century Questions.* Berkeley: University of California Press, 382–98.
Foucault, M. (1997) The Masked Philosopher. In: Rabinow, P. (ed.) *Michel Foucault – Ethics: Subjectivity and Truth.* Translated by Hurley, R. and others. London: Penguin, 321–28.
Foucault, M. (2006) *History of Madness.* Translated by Murphy, J. and Khalfa, J. London: Routledge.
Foucault, M. (2007) *Security, Territory, Population. Lectures at the Collège de France, 1977–1978.* Translated by Burchell, G. London: Palgrave.
Foucault, M. (2008) *The Birth of Biopolitics: Lectures at the Collège de France, 1978–1979.* Translated by Burchell, G. London: Palgrave.
Furedi, F. (2006) *Culture of Fear Revisited.* London: Continuum.
Fraad, H. (2000) Exploitation in the Labour of Love. In: Gibson-Graham, J-K., Resnick, S. and Wolff, R. (eds) *Class and its Others.* Minneapolis and London: University of Minnesota Press, 69–86.

Gallagher, C. (2006) *The Body Economic. Life, Death, and Sensation in Political Economy and the Victorian Novel.* Princeton: Princeton University Press.

Gatens, M. and Lloyd, G. (1999) *Collective Imaginings: Spinoza Past and Present.* London: Routledge.

Goleman, D. (1996) *Emotional Intelligence: Why It Can Matter More Than IQ.* London: Bloomsbury.

Goodman, S. (2009) *Sonic Warfare: Sound, Affect and the Ecology of Fear.* Cambridge, MA: MIT Press.

Greco, M. and Stenner, P. (2008) *Emotions: A Social Science Reader.* London: Routledge.

Green, A. (1977) Conceptions of affect. *International Journal of Psychoanalysis* 58, 129–56.

Greenhough, B. and Roe, E. (2010) Ethics, space, and somatic sensibilities: comparing relationships between scientific researchers and their human and animal experimental subjects. *Environment and Planning D: Society and Space* 29(1), 47–66.

Gregg, M. and Seigworth, G. (2010) *The Affect and Cultural Theory Reader.* Durham, NC and London: Duke University Press.

Gregory, D. (2007) Vanishing Points. In: Gregory, D. and Pred, A. (eds) *Violent Geographies: Fear, Terror, and Political Violence.* London: Routledge, 205–36.

Griffiths, P. (1997) *What Emotions Really Are: The Problem of Psychological Categories.* Chicago: University of Chicago Press.

Grossberg, L. (1997) *Dancing In Spite of Myself. Essays on Popular Culture.* Durham NC and London: Duke University Press.

Grossberg, L. (2010) *Cultural Studies in the Future Tense.* Durham NC and London: Duke University Press.

Grosz, E. (2005) *Time Travels: Feminism, Nature, Power.* Durham NC and London: Duke University Press.

Hage, G. (2003) *Against Paranoid Nationalism: Searching for Hope in a Shrinking Society.* Sydney: Pluto Press.

Hage, G. (2010) The affective politics of racial mis-interpellation. *Theory, Culture & Society* 27, 112–29.

Hall, S. (1980) Cultural studies: two paradigms. *Media, Culture and Society* 2, 57–72.

Hall, S. (1988) *The Hard Road to Renewal: Thatcherism and the Crisis of the Left.* London and New York: Verso.

Hall, S., Critcher, C., Jefferson, T., Clarke, J. and Roberts, B. (1978) *Policing the Crisis. Mugging, The State and Law and Order.* London: Palgrave Macmillan. Hamsson, T. (1976) *Living Through the Blitz.* London: Collins.

Harding, J. and Pribram, D. (2004) Losing our cool? Following Williams and Grossberg on emotions. *Cultural Studies* 18(6), 863–83.

Hardt, M. (2007) Foreword: What Affects Are Good For. In: Clough, P. with Halley, J. (eds) *The Affective Turn. Theorizing the Social.* Durham NC and London: Duke University Press, ix–xiii.

Hardt, M. and Negri, A. (2001) *Empire.* Cambridge, MA: Harvard University Press.

Hardt, M. and Negri, A. (2004) *Multitude. War and Democracy in the Age of Empire.* New York: Penguin.

Harrison, P. (2000) Making sense: embodiment and the sensibilities of the everyday. *Environment and Planning D: Society and Space* 18, 497–517.

Harrison, P. (2007) 'How shall I say it?' Relating the nonrelational. *Environment and Planning A* 39, 590–608.

Hemmings, C. (2005) Invoking affect: cultural theory and the ontological turn. *Cultural Studies* 19(5), 548–67.

Herman, E. (1995) *The Romance of American Psychology: Political Culture in the Age of Experts.* Berkeley: University of California Press.

Heron, W. (1965) Cognitive and Physiological Effects of Perceptual Isolation. In: Solomon, P. et al. (eds) *Sensory Deprivation: A Symposium Held at Harvard Medical School.* Cambridge, MA: Harvard University Press, 6–33.

Hewitt, K. (1994) 'When the great planes came and made ashes of our city': towards an oral geography of the disasters of war. *Antipode* 26, 1–34.

Hinchliffe, S. (2007) *Geographies of Nature. Societies, Environments, Ecologies.* London: Sage.

Hinkle, J. (1961) The Physiological State of the Interrogation Subject as it Affects Brain Function. In: Biderman, O. and Zimmmer, H. (eds) *Manipulation of Human Behavior.* New York: Wiley.

Hochschild, A. (1975) The Sociology of Feeling and Emotion: Selected Possibilities. In: Millman, M. and Kanter, R. (eds) *Another Voice.* New York: Anchor, 280–307.

Hochschild, A. (1979) Emotion work, feeling rules, and social structure. *The American Journal of Sociology* 85(3), 551–75.

Hochschild, A. (1983) *The Managed Heart: Commercialisation of Human Feeling.* California: University of California Press.

Hochschild, A. (2012) *The Outsourced Self: Intimate life in Market Times.* New York: Metropolitan Books.

Hocking, W. (1918) *Morale and Its Enemies.* New Haven: Yale University Press.

Hocking, W. (1941) The nature of morale. *The American Journal of Sociology* 47(3), 302–20.

Holloway, J. (2010) Legend-tripping in spooky space: ghost tourism and infrastructures of enchantment. *Environment and Planning D: Society and Space* 28(4), 618–37.

Hopkins, P. (2007) Young Muslim Men's Experiences of Local Landscapes after 11 September 2001. In: Aitchison, C., Hopkins, P. and Kwan, M. (eds) *Geographies of Muslim Identities: Gender, Diaspora and Belonging.* Farnham: Ashgate, 189–200.

Hunt, L. (2007) *Inventing Human Rights: A History.* New York: W.W. Norton & Co.

Isin, E. (2004) The neurotic citizen. *Citizenship Studies* 8(3), 217–35.

Jackson, P. (1989) *Maps of Meaning.* London and New York: Routledge.

Jayne, M., Valentine, G. and Holloway, S. (2010) Emotional, embodied and affective geographies of alcohol, drinking and drunkenness. *Transactions of the Institute of British Geographers* 35(4), 540–54.

Johns, F. (2005) Guantánamo Bay and the annihilation of the exception. *European Journal of International Law* 16, 613–35.

Jones, E., Woolven, R., Durodie, B. and Wessely, S. (2006) Public panic and morale: second world war civilian responses re-examined in the light of the current anti-terrorist campaign. *Journal of Risk Research* 9(1), 57–73.

Jones, R., Pykett, J. and Whitehead, M. (2011) Governing temptation: changing behaviour in an age of libertarian paternalism. *Progress in Human Geography* 35(4), 483–501.

Junger, E. (2004) *Storm of Steel*. Translated by Hofmann, M. London: Penguin.

Katz, J. (1999) *How Emotions Work*. Chicago: University of Chicago Press.

Kennett, L. (1982) *A History of Strategic Bombing*. New York: Charles Scribner's Sons.

Keynes, J. (1936) *The General Theory of Employment, Interest and Money*. London: Macmillan.

Khalili, L. (2012) *Time in the Shadows: Confinement in Counterinsurgencies*. Stanford: Stanford University Press.

Kinsley, S. (2010) Representing 'things to come': feeling the visions of future technologies. *Environment and Planning A* 42(11), 2771–90.

Kumar, A. (2000) Foreword: In Class. In: Gibson-Graham, J-K., Resnick, S. and Wolff, R. (eds) *Class and Its Others*. Minneapolis and London: University of Minnesota Press, vii – xiii.

Landis, J. (1941) Morale and civilian defense. *The American Journal of Sociology* 47(3), 331–39. Lasch, C. (1991) *The True and Only Heaven: Progress and Its Critics*. London: W.W. Norton & Co.

Latham, A. and McCormack, D. (2009) Thinking with images in non-representational cities: vignettes from Berlin. *Area* 41(3), 252–62.

Latour, B. (1988) *The Pasteurization of France*. Translated by Sheridan, A. and Law, J. Harvard: Harvard University Press.

Latour, B. (1993) *We Have Never Been Modern*. Translated by Porter, C. Harvard: Harvard University Press.

Latour, B. (1999) *Pandora's Hope: Essays on the Reality of Science Studies*. Harvard: Harvard University Press.

Latour, B. (2004a) Why has critique run out of steam? From matters of fact to matters of concern. *Critical Inquiry* 30, 225–48.

Latour, B. (2004b) How to talk about the body? The normative dimension of science studies. *Body and Society* 10(2–3), 205–30.

Latour, B. and Lépinay, V. (2009) *The Science of Passionate Interests: An Introduction to Gabriel Tarde's Economic Anthropology*. Chicago: Prickly Paradigm Press: Chicago.

Laurier, E. (2010) How to Feel Things with Words. In: Anderson, B. and Harrison, P. (eds) *Taking Place: Non-Representational Theories and Geography.* Farnham: Ashgate, 131–46.

Lazzarato, M. (1996) Immaterial Labor. In: Virno, P. and Hardt, M. (eds) *Radical Thought in Italy: A Potential Politics.* Minneapolis: University of Minnesota Press, 133–47.

Lazzarato, M. (2006) Life and the Living in the Societies of Control. In: Fuglsang, M. and Sorensen, B. (eds) *Deleuze and the Social.* Edinburgh: Edinburgh University Press, 171–90.

Lea, J. (2008) Retreating to nature: rethinking therapeutic landscapes. *Area* 80(1), 90–98.

Leduc, S. (2010) 'Confidence and the business cycle'. FRBSF Economic Letters. November. Available at: http://www.frbsf.org/publications/economics/letter/2010/el2010–35.html (last accessed March 2011).

Legg, S. (2005) Foucault's population geographies: classifications, biopolitics, and governmental spaces. *Population, Space and Place* 11(3), 137–56.

Legg, S. (2010) Assemblage/apparatus: using Deleuze and Foucault. *Area* 43(2), 128–33.

Legrenzi, P. and Umilta, C. (2011) *Neuromania: On the Limits of Brain Science.* Oxford: Oxford University Press.

Leys, R. (2007) *From Guilt to Shame. Auschwitz and After.* Princeton: Princeton University Press.

Leys, R. (2011) The turn to affect: a critique. *Critical Inquiry* 37, 434–72.

Lim, J. (2007) Queer Critique and the Politics of Affect. In: Browne, K., Lim, J. and Brown, G. (eds) *Geographies of Sexualities: Theory, Practices and Politics.* Farnham: Ashgate, 53–67.

Lim, J. (2010) Immanent politics: thinking race and ethnicity through affect and machinism. *Environment and Planning A* 42, 2393–409.

Lindeman, E. (1941) Recreation and morale. *The American Journal of Sociology* 47(3), 394–405.

Lindqvist, S. (2002) *A History of Bombing.* London: Granta.

Lingis, A. (2002) Murmurs of Life. In: Zournazi, M. (ed.) *Hope: New Philosophies for Change.* Sydney: Pluto Press, 22–42.

Lloyd, G. (1984) *The Man of Reason'.Male' and 'Female' in Western Philosophy.* Suffolk: Methuen.

Lorimer, H. (2005) Cultural geography: the busyness of being 'more-than-representational'. *Progress in Human Geography* 29(1), 83–94.

Lorimer, H. (2007) Cultural geography: worldly shapes, differently arranged. *Progress in Human Geography* 31(1), 89–100.

Lorimer, H. (2008) Cultural geography: nonrepresentational conditions and concerns. *Progress in Human Geography* 32(4), 551–9.Lupton, D. (1998) *The Emotional Self: A Sociocultural Exploration.* London: Sage.

McCormack, D. (2003) An event of geographical ethics in spaces of affect. *Transactions of the Institute of British Geographers* 4, 488–507.

McCormack, D. (2008) Engineering affective atmospheres: on the moving geographies of the 1897 Andree expedition. *Cultural Geographies* 15(4), 413–30.
McCoy, A. (2006) *A Question of Torture: CIA Interrogation from the Cold War to the War on Terror.* New York: Metropolitan Books, Henry Holt and Company.
McFarlane, C. (2009) Translocal assemblages: space, power and social movements. *Geoforum* 40(4), 561–67.
McLaine, I. (1979) *Ministry of Morale: Home Front Morale and the Ministry* of *Information in World War Two.* London: Allen and Unwin.
Maffesoli, M. (1996) *The Time of the Tribes.* Translated by Smith, D. London: Sage.
Mandel, E. (1978) *Late Capitalism.* Translated by De Bres, J. London: Verso.
Marcel, G. (1965) *Being and Having.* Translated by Black, A. London and Glasgow: Collins.
Marcel, G. (1967) Desire and Hope. Translated by Lawrence, N. In: Lawrence, N. and O'Connor, M. (eds) *Readings in Existential Phenomenology.* Englewood Cliffs, NJ: Prentice Hall, 277–86.
Marx, K. (1978) [1856] Speech at the Anniversary of the People's Paper. In: Tucker, R.C. (ed.) *The Marx-Engels Reader, Second Edition.* London: W.W. Norton & Co., 577–78.
Marx, K. (1995) [1867] *Capital: A Critique of Political Economy, Volume One.* Oxford and New York: Oxford University Press.
Massumi, B. (2002a) *Parables for the Virtual: Movement, Affect, Sensation.* Durham, NC: Duke University Press.
Massumi, B. (2002b) Introduction. Like a Thought. In: Massumi, B. (ed.) *A Shock to Thought. Expression after Deleuze and Guattari.* London: Routledge, xiii–xxxix.
Massumi B, (2002c) Navigating Movements. In: Zournazi, M. (ed.) *Hope: New Philosophies for Change.* Sydney: Pluto Press, 210–44.
Miller, J. (2012) Malls without stores. The affectual spaces of a Buenos Aires shopping mall. *Transactions of the Institute of British Geographers.* Online DOI:10.1111/j.1475–5661.2012.00553.x (last accessed 10 November 2012).
Mitropoulos, A. (2010) 'From precariousness to risk management and beyond'. Available at: http://archive.blogsome.com (last accessed 12 August 2011).
Moisi, D. (2009) *The Geopolitics of Emotion: How Cultures of Fear, Humiliation and Hope are Reshaping the World.* London: Bodley Head.
Mol, A. (2002) *The Body Multiple: Ontology in Medical Practice.* Durham, NC and London: Duke University Press.
Muehlebach, A. (2011) On affective labour in post-Fordist Italy. *Cultural Anthropology* 26(1), 59–82.
Muñoz, J. (2009) *Cruising Utopia: The Then and There of Queer Futurity.* New York: NYU Press.
Nealon, J. (2008) *Foucault Beyond Foucault: Power and its Intensifications Since 1984.* Stanford: Stanford University Press.

Neilson, B. and Rossiter, N. (2006) From precarity to precariousness and back again: labour, life, and unstable networks. Available at: http://www.variant.org.uk/pdfs/issue25/precarity.pdf (last accessed 10 March 2012).

Neilson, B. and Rossiter, N. (2008) Precarity as a political concept, or, Fordism as exception. *Theory, Culture & Society* 25(7–8), 51–72.

Ngai, S. (2005) *Ugly Feelings.* Cambridge, MA: Harvard University Press.

Nietzsche, F. (1986) *Human All Too Human, Volume 1.* Translated by Holland, R.J. Cambridge: Cambridge University Press.

Nunn, K. (1996) Personal hopefulness: a conceptual review of the relevance of the perceived future to psychology. *British Journal of Medical Psychology* 69, 227–45.

Office of Strategic Services Planning Group (1943) *Doctrine Regarding Rumours.* Available at: http://www.icdc.com/~paulwolf/oss/rumormanual2june1943.htm (last accessed 5 March 2007).

Ojakangas, M. (2005) Impossible dialogue on biopower: Agamben and Foucault. *Foucault Studies* 2, 5–28.

Ophir, A. (2007) The two-state solution: providence and catastrophe. *Journal of Homeland Security and Emergency Management* 4, (1), 1–44.

Orr, J. (2006) *Panic Diaries.* Durham, NC: Duke University Press.

O'Sullivan, S. (2001) The aesthetics of affect: thinking art beyond representation. *Angelaki: Journal of the Theoretical Humanities* 6(3), 125–35.

Otterman, M. (2007) *American Torture.* Melbourne: University of Melbourne Press.

Ó Tuathail, G. (2003) 'Just out looking for a fight': American affect and the invasion of Iraq. *Antipode* 35(5), 856–70.

Packard, V. (1957) *The Hidden Persuaders.* London: Penguin.

Pain, R. (2009) Globalized fear? Towards an emotional geopolitics. *Progress in Human Geography* 33, 466–86.

Papoulias, C. and Callard, F. (2010) Biology's gift: interrogating the turn to affect. *Body & Society* 16(1), 29–56.

Parisi, L. and Goodman, S. (2011) Mnemonic Control. In: Clough, P. (ed.) *Beyond Biopolitics: Essays on the Governance of Life and Death.* Durham, NC and London: Duke University Press.

Park, R. (1941) Morale and the news. *The American Journal of Sociology* 47(3), 360–77.

Parker, M. (ed.) (2002) *Utopia and Organization.* London: Blackwell.

Parse, R. (1999) *Hope: An International Human Becoming Perspective.* London: Jones and Bartlett Publishers International.

Paterson, M. (2006) Feel the presence: technologies of touch and distance. *Environment and Planning D: Society and Space* 24, 691–708.

Pfau, T. (2005) *Romantic Moods: Paranoia, Trauma, and Melancholy, 1790–1840.* Baltimore: Johns Hopkins University Press.

Philo, C. (2011) Discursive Life. In: Del Casino Jr, V. et al. (eds) *A Companion to Social Geography.* London: Blackwell, 362–84.

Pieper, J. (1994) *Hope and History.* Translated by Kipp, D. San Francisco: Ignatius.

Pile, S. (2010) Emotions and affect in recent human geography. *Transactions of the Institute of British Geographers* 35(1), 5–20.
Pope, A. (1941) The Importance of morale. *Journal of Educational Sociology* 15(4), 195–205.
Potamianou, A. (1997) *Hope: A Shield in the Economy of Borderline States.* London: Routledge.
Povinelli, E. (2011) *Economies of Abandonment.* Durham, NC and London: Duke University Press.
Preston, J. (2008) In the mi(d)st of. *Architectural Design* 6–11.
Probyn, E. (2000a) *CarnalAppetites: FoodSexIdentities.* London: Routledge.
Probyn, E. (2000b) Shaming theory, thinking dis-connections: feminism and reconciliation. In: Ahmed, S. et al. (eds) *Transformations: Thinking Through Feminism.* London: Routledge, 48–60.
Probyn, E. (2005) *Shame.* Minneapolis: University of Minnesota Press.
Puar, J. (2007) *Terrorist Assemblages: Homonationalism in Queer Times.*Durham, NC and London: Duke University Press.
Rabinow, P. (2007) *Marking Time: On the Anthropology of the Contemporary.* Princeton and London: Princeton University Press.
Ridout, N. and Schneider, R. (2012) Precarity and performance: an introduction. *TDR: The Drama Review* 56(4), 5–9.
Riley, D. (2005) *Impersonal Passion.* Durham, NC and London: Duke University Press.
Rodaway, P. (1994) *Sensuous Geographies.* London: Routledge.
Roe, E. (2006) Material connectivity, the immaterial and the aesthetic of eating practices: an argument for how genetically modified foodstuff becomes inedible. *Environment and Planning A* 38(3), 465–81.
Rose, G. (1993) *Feminism and Geography: The Limits of Geographical Knowledge* Minneapolis: University of Minnesota Press.
Rose, N. (1998) *Inventing Our Selves: Psychology, Power and Personhood* Cambridge University Press: Cambridge.
Saldanha, A. (2005) Trance and visibility at dawn: racial dynamics in Goa's rave tourism. *Social and Cultural Geography* 6, 707–21.
Saldanha, A, (2007) *Psychedelic White: Goa Trance and the Viscosity of Race.* Minneapolis: University of Minnesota.
Saldanha, A. (2010) Politics and difference. In: Anderson, B. and Harrison, P. (eds) *Taking Place: Non-Representational Theories and Geography.* Farnham: Ashgate, 283–302.
Sandbrook, D. (2011) *State of Emergency: The Way We Were: Britain, 1970–1974.* London: Penguin.
Scarry, E. (1985) *The Body In Pain.* Oxford: Oxford University Press.
Sedgwick, E.K. (2003) *Touching Feeling. Affect, Pedagogy, Performativity.* Durham, NC and London: Duke University Press.
Sedgwick, E.K. and Frank, A. (1995) Shame in the Cybernetic Fold: Reading Silvan Tomkins. In: Sedgwick, E.K. and Frank, A. (eds) *Shame and Its*

Sisters: A Silvan Tomkins Reader. Durham, NC and London: Duke University Press, 1–28.

Seigworth, G. (2006) Cultural Studies and Gilles Deleuze. In: Hall, G. and Birchall, C. (eds) *New Cultural Studies: Adventures in Theory.* Edinburgh: Edinburgh University Press, 107–27.

Seigworth, G. and Gregg, M. (2010) An Inventory of Shimmers. In: Gregg, M. and Seigworth, G. (eds) *The Affect Theory Reader.* Durham, NC and London: Duke University Press, 1–29.

Sennett, R. (1998) *The Corrosion of Character: The Personal Consequences of Work in the New Capitalism.* New York and London: W.W. Norton & Co.

Shaviro, S. (2009) *Without Criteria: Kant, Whitehead, Deleuze, and Aesthetics.* Cambridge, MA: MIT Press.

Shaviro, S. (2010) *Post Cinematic Affect.* Winchester: Zero Books.

Simpson, D. (1992) Raymond Williams: feeling for structures, voicing 'history'. *Social Text* 30, 9–26.

Simpson, P. (2008) Chronic everyday life: Rhythmanalysing street performance. *Social and Cultural Geography* 9(7), 807–29.

Sloterdijk, P. (2005) Foreword to the Theory of Spheres. In: Obanian, M. and Royaux, J. (ed.) *Cosmograms.* New York: Lukas and Sternberg, 223–40.

Smith, D.W. (1998) 'A Life of Pure Immanence'. Deleuze's Critique et Clinique project. In: *Gilles Deleuze: Essays Critical and Clinical.* Translated by Greco, M. and Smith, D. London: Verso, xi–lvi.

Southwood, I. (2011) *Non Stop Inertia.* Winchester: Zero Books.

Spinoza, B. (1910) *Ethics and the Treatise on the Correction of the Intellect.* Translated by Boyle, A. London: J.M. Dent & Sons.

Stern, D. (1998) *The Interpersonal World of the Infant.* London and New York: Karnac.

Stewart, K. (2007) *Ordinary Affects.* Durham, NC and London: Duke University Press.

Stiegler, B. (2010) *Taking Care of Youth and the Generations.* Translated by Barker, S. Stanford: Stanford University Press.

Sullivan, H. (1941) Psychiatric aspects of morale. *The American Journal of Sociology* 47(3), 277–301.

Swanton, D. (2010) Sorting bodies: race. Affect, and everyday multiculture in a mill town in northern England. *Environment and Planning A* 42, 2332–350.

Swyngedouw, E. (2010) Apocalypse forever?: Post-political populism and the spectre of climate change. *Theory, Culture and Society* 27, 213–32.

Taussig, M. (2002) Carnival of the Senses. In: Zournazi, M. (ed.) *Hope: New Philosophies for Change.* Sydney: Pluto Press, 42–63.

Terada, R. (2003) Feeling in Theory: Emotion After the 'Death of the Subject'. New Haven MA: Harvard University Press.

Terranova, T. (2012) Attention, economy and the brain. *Culture Machine* 13 (no pagination).

Thien, D. (2005) After or beyond feeling? A consideration of affect and emotion in geography. *Area* 37(4), 450–54.

Thornton, D. (2011) *Brain Culture: Neuroscience and Popular Media.* New York: Rutgers University Press.

Thrift, N. (2004a) Intensities of feeling: towards a spatial politics of affect. *Geografiska Annaler* 86, 57–78.

Thrift, N. (2004b) Summoning Life. In: Cloke, P., Crang, P. and Goodwin, M. (eds) *Envisioning Human Geography.* London: Arnold, 81–103.

Thrift, N. (2005) *Knowing Capitalism.* London: Sage.

Thrift, N. (2006) Overcome by Space: Reworking Foucault. In: Crampton, J. and Elden, S. (eds) *Space, Knowledge and Power: Foucault and Geography.* London: Ashgate, 53–8.

Thrift, N. (2007) *Non-Representational Theory: Space, Politics, Affect.* New York and London: Routledge.

Tiffany, D. (2000) *Toy Medium: Materialism and Modern Lyric.* Berkeley: University of California Press.

Tomkins, S. (1995) What are Affects? In: Kosofsky-Sedgwick, E. and Frank, A. *Shame and its Sisters: A Silvan Tomkins Reader.* Durham, NC and London: Duke University Press, 32–74.

Toscano, A. (2007) 'European Nihilism' and Beyond: Commentary by Alberto Toscano. In: Badiou, A. (2007) *The Century.* Translated by Toscano, A. London: Polity Press, 179–201.

Turkel, S. (2004) *Hope Dies Last.* London: The New Press.

Ulio, J. (1941) Military morale. *The American Journal of Sociology* 47(3), 321–30.

US Army/Marine Corps. (2006) *The US Army/Marine Corps Counterinsurgency Field Manual.* Chicago: University of Chicago Press.

US Strategic Bombing Survey (1947a) *A Detailed Study of the Effects of Area Bombing on Hamburg.* Washington: US Strategic Bombing Survey.

US Strategic Bombing Survey (1947b) *A Detailed Study of the Effects of Area Bombing on Wuppertal.* Washington: US Strategic Bombing Survey.

Van Creveld, M. (1991) *Technology and War.* Toronto: Maxwell Macmillan.

Virilio, P. and Lotringer, S. (1997) *Pure War.* New York: Semiotext(e).

Virno, P. (2004a) *A Grammar of the Multitude.* London: Semiotext(e).

Virno, P. (2004b) 'Creating a new public sphere, without the state'. Interview on libcom.org. Available at: http://libcom.org/library/creating-a-new-public-sphere-without-the-state-paolo-virno (last accessed 10 February 2010).

Waite, L. (2009) A place and space for a critical geography of precarity. *Geography Compass* 3(1), 412–33.

Waterworth, J. (2003) *A Philosophical Analysis of Hope.* London: Palgrave Macmillan.

Watson, G. (ed.) (1942) Civilian Morale. In: *Second Yearbook of the Society for the Psychological Study of Social Issues.* New York: Reynal and Hitchcock.

Weber, S. (2004) *Theatricality as Medium.* New York: Fordham University Press.

Weber, S. (2005) *Targets of Opportunity: On the Militarization of Thinking*. New York: Fordham University Press.
Widder, N. (2000) What's lacking in the lack: a comment on the virtual. *Angelaki: Journal of the Theoretical Humanities* 5(3), 117–38.
Wilkinson, I. (2001) *Anxiety in a Risk Society*. London: Routledge.
Williams, R. (1958) *Culture and Society,1780–1950*. Harmondsworth: Penguin Books.
Williams, R. (1961) *The Long Revolution*. Harmondsworth: Penguin Books.
Williams, R. (1966) *Modern Tragedy*. Stanford: Stanford University Press.
Williams, R. (1976) *Keywords. A Vocabulary of Culture and Society*. London: Harper Press.
Williams, R. (1977) *Marxism and Literature*. Oxford: Oxford University Press.
Williams, R. (1981) *Politics and Letters: Interviews with New Left Review*. London: Verso.
Williams, R. and Orrom, M. (1954) *A Preface to Film*. London: Film Drama.
Williams, S. (2001) *Emotions and Social Theory*. London: Sage.
Wilson, E. (1998) *Neural Geographies: Feminism and the Microstructure of Cognition*. New York: Routledge.
Wilson, H. (2011) Passing propinquities in the multicultural city: the everyday encounters of bus passengering. *Environment and Planning A* 43(3), 634–49.
Woodward, K. (2009) *Statistical Panic: Cultural Politics and Poetics of the Emotions*. Durham, NC and London: Duke University Press.
Woodward, K. and Lea, J. (2010) Geographies of Affect. In: Smith, S. et al. (eds) *The Sage Handbook of Social Geographies*. London: Sage, 154–75.
Woodyer, T. and Geoghegan, H. (2013) (Re)enchanting geography? The nature of being critical and the character of critique in human geography. *Progress in Human Geography* 37(2), 195–214.
Wylie, J. (2005) A single day's walking: narrating self and landscape on the south west coast path. *Transactions of the Institute of British Geographers* 30(2), 234–47.
Wylie, J. (2006) Depths and folds: on landscape and the gazing subject. *Environment and Planning D: Society and* Space 24(4), 519–35.
Wylie, J. (2009) Landscape, absence and the geographies of love. *Transactions of the Institute of British Geographers* 34(3), 275–89.
Zumthor, P. (2006) *Atmospheres: Architectural Environments – Surrounding Objects*. Berlin: Birkhauser Verlag AG.

Index

Index of Names